S0-AHN-543

Handbook
of
Writing
for the
Mathematical
Sciences

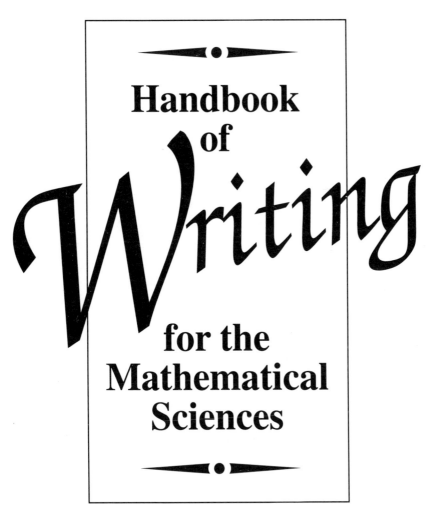

Handbook
of
Writing
for the
Mathematical
Sciences

NICHOLAS J. HIGHAM
University of Manchester

siam®

Society for Industrial and Applied Mathematics
Philadelphia   1993

WITHDRAWN
LORETTE WILMOT LIBRARY
NAZARETH COLLEGE

Library of Congress Cataloging-in-Publication Data

Higham, N. J.
    Handbook of writing for the mathematical sciences/ Nicholas J.
Higham.
        p. cm.
    Includes bibliographical references and index.
    ISBN 0-89871-314-5
    1. Mathematics—Authorship. 2. Technical writing. I. Title.
QA42.H54 1993
808' .06651—dc20   93-12791

All rights reserved. Printed in the United States of America. No part of this book may be reproduced, stored, or transmitted in any manner without the written permission of the Publisher. For information, write the Society for Industrial and Applied Mathematics, 3600 University City Science Center, Philadelphia, Pennsylvania 19104-2688.

Copyright 1993 by the Society for Industrial and Applied Mathematics
Second printing 1994.

siam₀ is a registered trademark.

Contents

Preface

In this book I aim to describe most of what a scientist needs to know about mathematical writing. Although the emphasis is strongly on *mathematical* writing, many of the points and issues I discuss are relevant to *scientific* writing in general. My main target audience is graduate students. They often have little experience or knowledge of technical writing and are daunted by the task of writing a report or thesis. The advice given here reflects what I have learned in the ten years since I wrote my first research report as a graduate student and describes what I would have liked to know as I started to write that first report. I hope that as well as being a valuable resource for graduate students, this book will also be of use to practising scientists.

The book has grown out of notes for a short lecture course on mathematical writing that I gave at the University of Manchester in May 1992. As I prepared the course I realized that, although several excellent articles and books on mathematical writing are available (notably those by Halmos (1970) [91], Gillman (1987) [81] and Knuth, Larrabee and Roberts (1989) [128]), none functions as a comprehensive handbook that can be both read sequentially and used as a reference when questions about mathematical writing arise. I have attempted to provide such a handbook. (I hope that the comment of one journal referee, "This paper fills a much needed gap in the literature", is not applicable to this book!)

As well as covering standard topics such as English usage, the anatomy of a research paper, and revising a draft, I examine in detail four topics that are usually discussed only briefly, if at all, in books on technical writing.

- The whole publication process, from submission of a manuscript to its appearance as a paper in a journal.

- Writing when English is a foreign language.

- How to write slides for a talk.

- The use of computers in writing and research. In particular, I discuss modern practices such as computerized typesetting in TeX, the use of computer tools for indexing and checking spelling and style, and electronic mail and ftp (file transfer protocol).

An important feature of the book is that many examples are given to illustrate the ideas and principles discussed. In particular, Chapter 7 contains a collection of extracts from the mathematics and computer science literature, with detailed comments on how each extract can be improved. In writing the book I have been helped and influenced by many people. Several people read the entire manuscript at one or more of its various stages, offered constructive suggestions, encouragement and advice, and made sure I said what I meant and meant what I said. They are

> Ian Gladwell, Des Higham, Doris Higham, Nil Mackey, Fred Schneider, Pete (G. W.) Stewart and Nick Trefethen.

Other people who have read portions of the book and have given help, suggestions or advice are

> Carl de Boor, David Carlisle, Valérie Frayssé, Paul Halmos, Bo Kågström, Philip Knight, Sven Leyffer, Steve Mackey, Volker Mehrmann, June O'Brien, Pythagoras Papadimitriou, Beresford Parlett, Nigel Ray, Stephan Rudolfer, Bob Sandling, Zdenek Strakos, Gil Strang, Charlie Van Loan, Rossana Vermiglio, Joan Walsh, Barry White, and Yuanjing Xu.

I thank all these people, together with many others who have answered my questions and made suggestions. In researching the contents of the book I was inspired by many of the references listed in the bibliography, and learned a lot from them. I acknowledge the Nuffield Foundation for the support of a Nuffield Science Research Fellowship, during the tenure of which I wrote much of the book. Last, but not least, I thank the SIAM staff for their help and advice in the production of the book—in particular, Susan Ciambrano, Beth Gallagher and Tricia Manning. I would be happy to receive comments, notification of errors, and suggestions for improvement, which I will collect for inclusion in a possible second edition.

This book was typeset in LaTeX using the **book** document style with G. W. Stewart's **jeep** style option. I prepared the references with BibTeX and the index with MakeIndex. I used text editors Qedit (Semware) and GNU Emacs (Free Software Foundation), and checked spelling with PC-Write (Quicksoft).

Manchester
December 1992

Nicholas J. Higham

Chapter 1
General Principles

Peanuts reprinted by permisson of UFS, Inc.

Good writing ... is clear thinking made visible.
— AMBROSE BIERCE, *Write it Right:*
A Little Blacklist of Literary Faults (1937)

A writer needs three qualities:
creativity, originality, clarity and a good short term memory.
— DESMOND J. HIGHAM, *More Commandments of Good Writing* (1992)

1

Writing helps you to learn. Writing is not simply a task to be done once research or other preparation is completed—it can be an integral part of the work process. Baker explains it well [10]:

> In writing, you clarify your own thoughts, and strengthen your conviction. Indeed, you probably grasp your thoughts for the first time. Writing is a way of thinking. Writing actually creates thought, and generates your ability to think: you discover thoughts you hardly knew you had, and come to know what you know. You learn as you write.

Writing brings out gaps in your understanding, by forcing you to focus on steps in your thinking that you might otherwise skip. (In this respect, writing is similar to explaining your ideas to a colleague.) Writing down partial research results as your work progresses can reveal structure and suggest further areas of development. Zinsser has written a delightful book that explores the idea of writing to learn [224].

Good writing reflects clear thinking. It is very hard for a woolly thinker to produce clear prose. But clear thinking is only a necessary, not a sufficient, condition for good writing. Clear thinking leads to good organization, a vital ingredient of technical writing. A well-organized piece is much easier to write than a badly organized one. If you find a particular piece difficult to write it may be because you have not found the right structure to express your ideas.

Writing is difficult! Zinsser [225] says "It's one of the hardest things that people do." It is often difficult to get started. One solution, which works for some, is to force yourself to write something, however clumsy it may be, for it is often easier to modify something you have written previously than to compose from scratch.

The most fundamental tenet of technical writing is to keep your prose simple and direct. Much of written English is unnecessarily complicated. In writing up your research you are aiming at a relatively small audience, so it is important not to alienate part of it with long-winded or imprecise text. English may not be the first language of many of your readers—they, particularly, will appreciate plain writing. Aim for economy of words. Early drafts can usually be reduced in length substantially with consequent improvements in readability (see Chapter 7).

Probably the best way to improve your writing skills is to receive, and learn from, constructive criticism. Ask a colleague to read and comment on your writing. Another reader can often find errors and suggest improvements that you miss because of your familiarity with the work. Criticism can be difficult to take, but it should be welcomed; it is a privilege to have someone else take the time to comment on your writing.

Another way to improve your writing is to read as much as you can, always with a critical eye. In particular, I recommend perusal of some of the following mathematical books. They are by excellent writers, each of whom has his own distinctive style (this selection is inevitably biased towards my own area of research):

FORMAN S. ACTON (1970), *Numerical Methods That Work* [3].

ALBERT H. BEILER (1966), *Recreations in the Theory of Numbers* [15].

DAVID M. BURTON (1980), *Elementary Number Theory* [32].

GENE H. GOLUB AND CHARLES F. VAN LOAN (1989), *Matrix Computations* [84].

PAUL R. HALMOS (1982), *A Hilbert Space Problem Book* [95].

DONALD E. KNUTH (1973–1981), *The Art of Computer Programming* [121].

BERESFORD N. PARLETT (1980), *The Symmetric Eigenvalue Problem* [160].

G. W. STEWART (1973), *Introduction to Matrix Computations* [195].

GILBERT STRANG (1986), *Introduction to Applied Mathematics* [197].

Also worth studying are papers or books that have won prizes for expository writing in mathematics. Appendix E lists winners of the Chauvenet Prize, the Lester R. Ford Award, the George Polya Award, the Carl B. Allendoerfer Award, the Beckenbach Book Prize and the Merten M. Hasse Prize.

Chapter 2
Writer's Tools and Recommended Reading

I use three dictionaries almost every day.

— JAMES A. MICHENER, *Writer's Handbook* (1992)

*The purpose of an ordinary dictionary is simply
to explain the meaning of the words ....
The object aimed at in the present undertaking is exactly the converse of this:
namely,—The idea being given, to find the word, or words,
by which that idea may be most fitly and aptly expressed.*

— PETER MARK ROGET, *Thesaurus of English Words and Phrases* (1852)

Apart from pen, paper and keyboard, the most valuable tool for a writer is a dictionary. It is used to check spelling, shades of meaning, and usage. Since the English language is constantly changing, if you refer to only one dictionary it is best to use a recent edition—a dictionary from twenty years ago will not fully reflect current vocabulary and usage. It is worth noting, however, that the $(n+1)$st edition of a dictionary is not always universally regarded as being better than the nth. For example, Zinsser [225] clings to his second edition of Webster's Dictionary "because the Third Edition is quite permissive."

The most authoritative dictionary is the *Oxford English Dictionary* (OED) [157]. It was originally published in parts between 1884 and 1928, and a four volume supplement was produced from 1972–1986. A twenty volume second edition of the dictionary was published in 1989; it defines more than half a million words, using 2.4 million illustrative quotations. The OED traces the history of words from around 1150. In 1992 a compact disc (CD Rom) version of the OED was published. It contains the full text of the printed version (at about a third of the price) and the accompanying software includes powerful search facilities. The major competitor to the OED is *Webster's Third New International Dictionary*, which was published in the United States in 1961 and has had three supplements.

For everyday use the OED and Webster's Third are too unwieldy and too thorough, so a more concise dictionary is needed. The *Concise Oxford Dictionary* (COD) [158] is now in its eighth edition (1990). It is the favourite of many, and is suitable for American use, as American spellings and usages are included. (The COD was my main dictionary of reference in writing this book.) Other important dictionaries include, from the United States, *The American Heritage Dictionary* [109], *The Random House College Dictionary* [176], *Webster's Ninth New Collegiate Dictionary* [147], *Webster's New World Dictionary* [172], and, from Britain, the *Chambers Concise Dictionary* [40], the *Collins English Dictionary* [45], and the *Longman Dictionary of the English Language* [137]. A notable feature of the Longman Dictionary is the many usage notes scattered throughout the text. Several of these dictionaries include entries for mathematical terms such as Cauchy sequence, Galois theory, and Rolle's theorem!

Annotated lists of dictionaries and usage guides are given by Stainton in [188] and [189]. Comparisons and analyses of dictionaries are also given by Quirk and Stein [175, Chap. 11] and Burchfield [31].

Specialized dictionaries can also be useful to the mathematical writer. There are many dictionaries of mathematics, one example being the Penguin dictionary [49], which is small and inexpensive yet surprisingly thorough.

The synonyms provided in a thesaurus can be helpful in your search for an elusive word or a word with the right connotation. *Roget's Thesaurus* is the classic one. It is available in various editions, the first published in 1852. One edition is available in electronic form—see §10.6. The words in Roget's Thesaurus are traditionally arranged according to the ideas they express, instead of alphabetically, though a Roget's Thesaurus is now available in dictionary form. Thesauri are produced by all the major publishers of dictionaries.

Every writer should own and read a guide to English usage. One of the most accessible is *The Elements of Style* by Strunk and White [198]. Zinsser [225] says this is "a book that every writer should read at least once a year," and, as if following this advice, Luey [138] says "I read it once a year without fail." Fowler's *Dictionary of Modern English Usage* [63] is a much longer and more detailed work, as is its predecessor, *The King's English*, by the Fowler brothers [64]. A favourite of mine is the revision [221] of the 1962 *Current English Usage* by Wood. Gowers's influential *Complete Plain Words* [87] stems from his *Plain Words* of 1948, which was written to improve the use of English in the British civil service. Partridge's *Usage and Abusage* [161] is another valuable guide, this one in dictionary form. Carey's *Mind the Stop* [39] provides an easy-to-read and illuminating explanation of punctuation.

Bryson's *Dictionary of Troublesome Words* [29] offers practical, witty advice on usage, while Safire [181] presents fifty "fumblerules" (mistakes that call attention to the rules) accompanied by pithy explanatory essays. The books *On Newspaper Style* and *English our English* by Waterhouse [215], [216] make fascinating and informative reading, though they are hard to use for reference since they lack an index; [215] is a standard handbook for journalists, but is of much wider interest. Baker's *Practical Stylist* [10] is a widely used course text on writing; it has thorough discussions of usage, style and revision and gives many illustrative examples. Day's *Scientific English* [52] contains general advice on grammar and usage, with particular reference to English in scientific writing. Perry's *The Fine Art of Technical Writing* [165] offers selective, practical advice on the psychology, artistry and technique of technical writing, which the author defines as "all writing other than fiction". In *Miss Thistlebottom's Hobgoblins* [21] Bernstein provides an antidote for those brainwashed by over-prescriptive usage guides, in the form of letters to his (fictional) English schoolteacher.

The Chicago Manual of Style [212], first published in 1906, is a long and comprehensive guide to book production, style and printing. It is the standard reference for authors and editors in many organizations. It includes chapters on typesetting mathematics and preparing bibliographies and indexes. Turabian's *A Manual for Writers of Term Papers, Theses, and Dissertations* [207], first published in 1937, is based on the guidelines in *The Chicago Manual of Style* but its aim is more limited, as defined in the title, so it does not discuss bookmaking and copy editing. *Words into Type* [184] is another thorough guide for authors and editors, covering manuscript and index preparation, copy editing style, grammar, typographical style and the printing process. Other valuable references on editing, copy editing and proofreading are *Hart's Rules* [99], which describes the house style of Oxford University Press; Butcher's *Copy-Editing* [33], which is regarded as the standard British work on copy editing; Eisenberg's *Guide to Technical Editing* [58]; O'Connor's *How to Copyedit Scientific Books and Journals* [151]; Stainton's *The Fine Art of Copyediting* [189]; and Tarutz's *Technical Editing* [203]. O'Connor has also written a book about how to edit and manage scientific books and journals [150].

Several guides to mathematical writing are available. Halmos's essay "How to Write Mathematics" [91] is essential reading for every mathematician; it contains much sound advice not found elsewhere. Halmos's "automathography" [97] includes much insight into mathematical writing, editing and refereeing; it begins with the sentence "I like words more than numbers, and I always did." Transcripts of a lecture course called "Mathematical Writing" that was given by Knuth in 1987 at Stanford are collected in *Mathematical Writing* [128], which I highly recommend. This manual contains many anecdotes and insights related by Knuth and his guest lecturers, including Knuth's battle with the copy editors at Scientific American and his experiences in writing the book *Concrete Mathematics* [88]. Other very useful guides are Flanders's guide [60] for authors who write in the journal *American Mathematical Monthly*; Gillman's booklet *Writing Mathematics Well* [81] on preparing manuscripts for Mathematical Association of America journals; Steenrod's essay "How to Write Mathematics" [191]; and Swanson's guide *Mathematics Into Type* [202] for mathematical copy editors and authors. Knuth's book on TEX [124] contains much general advice on how to typeset mathematics, and an old guide to this subject is *The Printing of Mathematics* [41].

Guides to writing in other scientific disciplines often contain much that is relevant to the mathematical writer; an example is the book by Pechenik [163], which is aimed at biology students. General guides to scientific writing that I recommend are those by Barrass [11], [12], Cooper [46], Ebel, Bliefert and Russey [57], Kirkman [119], O'Connor [152] (this is a revised and extended version of an earlier book by O'Connor and Woodford [153]), and Turk and Kirkman [209]. Particularly informative and pleasant to read are Booth's *Communicating in Science: Writing and Speaking* [27] and Day's *How to Write and Publish a Scientific Paper* [51].

The journal *IEEE Transactions on Professional Communication* publishes papers on many aspects of technical communication, including how to write papers and give talks. A selection of 63 papers from this and other journals is collected in *Writing and Speaking in the Technology Professions: A Practical Guide* [14].

How to Do It [135] contains 47 chapters that give advice for medical doctors, but many of them are of general interest to scientists. Chapter titles include "Write a Paper", "Referee a Paper", "Attract the Reader", "Review a Book", "Use an Overhead Projector", and "Apply for a Research Grant". Many of the chapters originally appeared in the *British Medical Journal*.

Van Leunen's *A Handbook for Scholars* [213] is a unique and indispensable guide to the mechanics of scholarly writing, covering reference lists, quotations, citations, footnotes and style. This is the place to look if you want to know how to prepare a difficult reference or quotation (what date to list for a reprint of a work from a previous century, or how to punctuate a quotation placed in mid-sentence). There is also an appendix on how to prepare a CV. Luey's *Handbook for Academic Authors* [138] offers much useful advice to the writer of an academic book.

The three books by Zinsser [223], [224], [225] are highly recommended; all are informative and beautifully written. In *Writing with a Word Processor* [223] Zinsser summarizes his experience in moving to a computer from his trusty typewriter. His book *Writing to Learn* contains chapters on "Writing Mathematics" and "Writing Physics and Chemistry"; they explain how writing can be used in the teaching of these subjects and give examples of good writing. Michener's *Writer's Handbook* [148] provides insight into how this prodigious writer works. The reader is led through the development of parts of two of Michener's books (one fiction, one nonfiction), from early drafts to proofs to the published versions. Thirty-one essays discussing how writing is being used to teach mathematics in undergraduate courses are contained in *Using Writing to Teach Mathematics* [193].

Mitchell [149] gives hints on writing with a computer, with good examples of how to revise drafts.

Valuable insight into the English language—its history, its eccentricities, and its uses—is provided by Bryson [30], Crystal [48] and Potter [171].

A useful source for examples of expository mathematical writing is the annotated bibliography of Gaffney and Steen [66], which contains more than 1100 entries.

Finally, Pemberton's book *How to Find Out in Mathematics* [164] tells you precisely what the title suggests. It includes information on mathematical dictionaries (including interlingual ones) and encyclopedias, mathematical histories and biographies, and mathematical societies, periodicals and abstracts. Although it appeared in 1969, the book is still worth consulting.

Even books on writing can have occasional lapses. One of the books above advises *There is no reason you absolutely must start at the beginning,* which surely reads better without the word *absolutely.*

Chapter 3
Mathematical Writing

Suppose you want to teach the "cat" concept to a very young child.
Do you explain that a cat is a relatively small,
primarily carnivorous mammal with retractile claws,
a distinctive sonic output, etc.?
I'll bet not.
You probably show the kid a lot of different cats,
saying "kitty" each time, until it gets the idea.
To put it more generally,
generalizations are best made by abstraction from experience.
— RALPH P. BOAS, *Can We Make Mathematics Intelligible?* (1981)

A good notation should be unambiguous, pregnant, easy to remember;
it should avoid harmful second meanings, and
take advantage of useful second meanings;
the order and connection of signs should
suggest the order and connection of things.
— GEORGE POLYA, *How to Solve It* (1957)

My most nearly immortal contributions to mathematics are
an abbreviation and a typographical symbol.
I invented "iff," for "if and only if"—but
I could never believe that I was really its first inventor . . . .
The symbol is definitely not my invention—it appeared in
popular magazines (not mathematical ones) before I adopted it,
but, once again, I seem to have introduced it into mathematics.
It is the symbol that sometimes looks like □,
and is used to indicate an end, usually the end of a proof.
It is most frequently called the "tombstone,"
but at least one generous author referred to it as the "halmos."
— PAUL R. HALMOS, *I Want to be a Mathematician:*
An Automathography in Three Parts (1985)

11

The mathematical writer needs to be aware of a number of matters specific to mathematical writing, ranging from general issues, such as choice of notation, to particular details, such as how to punctuate mathematical expressions. In this chapter I begin by discussing some of the general issues and then move on to specifics.

3.1. What is a Theorem?

What are the differences between theorems, lemmas, and propositions? To some extent, the answer depends on the context in which a result appears. Generally, a *theorem* is a major result that is of independent interest. The proof of a theorem is usually nontrivial. A *lemma*[1] is an auxiliary result—a stepping stone towards a theorem. Its proof may be easy or difficult. A straightforward and independent result that is worth encapsulating but that does not merit the title of a theorem may also be called a lemma. Indeed, there are some famous lemmas, such as the Riemann–Lebesgue Lemma in the theory of Fourier series and Farkas's Lemma in the theory of constrained optimization. Whether a result should be stated formally as a lemma or simply mentioned in the text depends on the level at which you are writing. In a research paper in linear algebra it would be inappropriate to give a lemma stating that the eigenvalues of a symmetric positive definite matrix are positive, as this standard result is so well known; but in a textbook for undergraduates it would be sensible to formalize this result.

It is not advisable to label all your results theorems, because if you do so you miss the opportunity to emphasize the logical structure of your work and to direct attention to the most important results. If you are in doubt about whether to call a result a lemma or a theorem, call it a lemma.

The term *proposition* is less widely used than lemma and theorem and its meaning is less clear. It tends to be used as a way to denote a minor theorem. Lecturers and textbook authors might feel that the modest tone of its name makes a proposition appear less daunting to students than a theorem. However, a proposition is not, as one student thought, "a theorem that might not be true."

A *corollary* is a direct or easy consequence of a lemma, theorem or proposition. It is important to distinguish between a corollary, which does not imply the parent result from which it came, and an extension or generalization of a result. Be careful not to over-glorify a corollary by failing to label it as such, for this gives it false prominence and obscures the role of the parent result.

---

[1]The plural of lemma is lemmata, or, more commonly, lemmas.

How many results are formally stated as lemmas, theorems, propositions or corollaries is a matter of personal style. Some authors develop their ideas in a sequence of results and proofs interspersed with definitions and comments. At the other extreme, some authors state very few results formally. A good example of the latter style is the classic book *The Algebraic Eigenvalue Problem* [220] by Wilkinson, in which only four titled theorems are given in 662 pages. As Boas [26] notes, "A great deal can be accomplished with arguments that fall short of being formal proofs."

A fifth kind of statement used in mathematical writing is a *conjecture*— a statement that the author thinks may be true but has been unable to prove or disprove. The author will usually have some strong evidence for the veracity of the statement. A famous example of a conjecture is the Goldbach conjecture (1742), which states that every even number greater than 2 is the sum of two primes; this is still unproved. One computer scientist (let us call him Alpha) joked in a talk "This is the Alpha and Beta conjecture. If it turns out to be false I would like it to be known as Beta's conjecture." However, it is not necessarily a bad thing to make a conjecture that is later disproved: identifying the question that the conjecture aims to answer can be an important contribution.

A *hypothesis* is a statement that is taken as a basis for further reasoning, usually in a proof—for example, an induction hypothesis. Hypotheses that stand on their own are uncommon; two examples are the Riemann hypothesis and the continuum hypothesis.

3.2. Proofs

Readers are often not very interested in the details of a proof but want to know the outline and the key ideas. They hope to learn a technique or principle that can be applied in other situations. When readers do want to study the proof in detail they naturally want to understand it with the minimum of effort. To help readers in both circumstances, it is important to emphasize the structure of a proof, the ease or difficulty of each step, and the key ideas that make it work. Here are some examples of the sorts of phrases that can be used (most of these are culled from proofs by Parlett in [160]).

> The aim/idea is to
> Our first goal is to show that
> Now for the harder part.
> The trick of the proof is to find
> ... is the key relation.
> The only, but crucial use of ... is that

To obtain ... a little manipulation is needed.
The essential observation is that

When you omit part of a proof it is best to indicate the nature and length of the omission, via phrases such as the following.

It is easy/simple/straightforward to show that
Some tedious manipulation yields
An easy/obvious induction gives
After two applications of ... we find

You should also strive to keep the reader informed of where you are in the proof and what remains to be done. Useful phrases include

First, we establish that
Our task is now to
Our problem reduces to
It remains to show that
We are almost ready to invoke
We are now in a position to
Finally, we have to show that

The end of a proof is often marked by the halmos symbol □ (see the quote on page 11). Sometimes the abbreviation QED (Latin: quod erat demonstradum = which was to be demonstrated) is used instead.

3.3. The Role of Examples

A pedagogical tactic that is applicable to all forms of technical writing (from teaching to research) is to discuss specific examples before the general case. It is tempting, particularly for mathematicians, to adopt the converse approach, but beginning with examples is often the more effective way to explain (see Boas's article [26] and the quote from it at the beginning of this chapter, a quote that itself illustrates this principle!).

A good example of how to begin with a specific case is provided by Strang in *Introduction to Applied Mathematics* [197]. He begins Chapter 1 as follows.

The simplest model in applied mathematics is a system of linear equations. It is also by far the most important, and we begin this book with an extremely modest example:

$$2x_1 + 4x_2 = 2$$
$$4x_1 + 11x_2 = 1.$$

After some further introductory remarks, Strang goes on to study in detail both this 2×2 system and a particular 4×4 system. General $n \times n$ matrices do not appear for several pages.

Another example is provided by Watkins's *Fundamentals of Matrix Computations* [217]. Whereas most linear algebra textbooks introduce Gaussian elimination (for general matrices) before discussing Cholesky factorization for symmetric positive definite matrices, Watkins reverses the order, giving the more specific but algorithmically more straightforward method first.

An exercise in a textbook is a form of example. I saw a telling criticism in one book review that complained "The first exercise in the book was pointless, so why do the others?" To avoid such criticism, it is important to choose exercises and examples that have a clear purpose and illustrate a point. The first few exercises and examples should be among the best, to gain the reader's confidence. The same reviewer complained of another book that "it hides information in exercises and contains exercises that are too difficult." Whether such criticism is valid depends on your opinion of what the key issues to be transmitted to the reader are and on the level of the readership. Again, it helps to bear such potential criticism in mind when you write.

3.4. Definitions

Three questions to be considered when formulating a definition are "why?", "where?" and "how?" First, ask yourself why you are making a definition: is it really necessary? Inappropriate definitions can complicate a presentation and too many can overwhelm a reader, so it is wise to imagine yourself being charged a large sum for each one. Instead of defining a square matrix A to be contractive with respect to a norm $\| \cdot \|$ if $\|A\| < 1$, which is not a standard definition, you could simply say "A with $\|A\| < 1$" whenever necessary. This is easy to do if the property is needed on only a few occasions, and saves the reader having to remember what "contractive" means. For notation that is standard in a given subject area, judgement is needed to decide whether the definition should be given. Potential confusion can often be avoided by using redundant words. For example, if $\rho(A)$ is not obviously the spectral radius of the matrix A you can say "the spectral radius $\rho(A)$".

The second question is "where?" The practice of giving a long sequence of definitions at the start of a work is not recommended. Ideally, a definition should be given in the place where the term being defined is first used. If it is given much earlier, the reader will have to refer back, with a possible loss of concentration (or worse, interest). Try to minimize the distance between a definition and its place of first use.

It is not uncommon for an author to forget to define a new term on its first occurrence. For example, Steenrod uses the term "grasshopper reader" on page 6 of his essay on mathematical writing [191], but does not define it until it occurs again on the next page.

To reinforce notation that has not been used for a few pages you may be able to use redundancy. For example, "The optimal steplength α^* can be found as follows." This implicit redefinition either reminds readers what α^* is, or reassures them that they have remembered it correctly.

Finally, how should a term be defined? There may be a unique definition or there may be several possibilities (a good example is the term M-matrix, which can be defined in at least fifty different ways [19]). You should aim for a definition that is short, expressed in terms of a fundamental property or idea, and consistent with related definitions. As an example, the standard definition of a normal matrix is a matrix $A \in \mathbf{C}^{n \times n}$ for which $A^*A = AA^*$ (where $*$ denotes the conjugate transpose). There are at least 70 different ways of characterizing normality [90], but none has the simplicity and ease of use of the condition $A^*A = AA^*$.

By convention, *if* means *if and only if* in definitions, so do not write "The graph G is connected if and only if there is a path from every node in G to every other node in G." Write "The graph G is connected if there is a path from every node in G to every other node in G" (and note that this definition can be rewritten to omit the symbol G). It is common practice to italicize the word that is being defined: "A graph is *connected* if there is a path from every node to every other node." This has the advantage of making it perfectly clear that a definition is being given, and not a property. This emphasis can also be imparted by writing "A graph is defined to be connected if ... ," or "A graph is said to be connected if ...."

If you have not done so before, it is instructive to study the definitions in a good dictionary. They display many of the attributes of a good mathematical definition: they are concise, precise, consistent with other definitions, and easy to understand.[2]

3.5. Notation

Consider the following extract.

> Let $\widehat{H_k} = Q_k^H \widetilde{H_k} Q_k$, partition $X = [X_1,\ X_2]$ and let $\mathcal{X} = \text{range}(X_1)$. Let U^* denote the nearest orthonormal matrix to

---

[2]Dictionary definitions can also be incorrect. One general dictionary offers this definition of eigenvector: a vector that in one dimension under a given rotational, reflectional, expanding, or shrinking operation becomes a number that is a multiple of itself.

X_1 in the 2-norm.

These two sentences are full of potentially confusing notation. The distinction between the hat and the tilde in $\widehat{H}_k$ and $\widetilde{H}_k$ is slight enough to make these symbols difficult to distinguish. The symbols $\mathcal{X}$ and X are also too similar for easy recognition. Given that $\mathcal{X}$ is used, it would be more consistent to give it a subscript 1. The name H_k is unfortunate, because H is being used to denote the conjugate transpose, and it might be necessary to refer to $\widetilde{H}_k^H$! Since A^* is a standard synonym for A^H, the use of a superscripted asterisk to denote optimality is confusing. As this example shows, the choice of notation deserves careful thought. Good notation strikes a balance among the possibly conflicting aims of being readable, natural, concise, logical and aesthetically pleasing. As with definitions, the amount of notation should be minimized.

Although there are 26 letters in the alphabet and nearly as many again in the Greek alphabet, our choice diminishes rapidly when we consider existing connotations. Traditionally, ϵ and δ denote small quantities, i, j, k, m and n are integers (or i or j the imaginary unit), λ is an eigenvalue and π and e are fundamental constants; π is also used to denote a permutation. These conventions should be respected. But by modifying and combining eligible letters we widen our choice. Thus γ and A yield, for example, $\hat{A}$, $\overline{A}$, $\widetilde{A}$, A', γ_A, A_γ, $\mathbf{A}$, $\mathcal{A}$.

Particular areas of mathematics have their own notational conventions. For example, in numerical linear algebra lower case Greek letters represent scalars, lower case roman letters represent column vectors, and upper case Greek or roman letters represent matrices. This convention was introduced by Householder [110].

In his book on the symmetric eigenvalue problem [160], Parlett uses the symmetric letters A, H, M, T, U, V, W, X, Y to denote symmetric matrices and the symmetric Greek letters Λ, Θ, Φ, Δ to denote diagonal matrices. Actually, the roman letters printed above are not symmetric because they are slanted, but Parlett's book uses a sans serif mathematics font that yields the desired symmetry. Parlett uses this elegant, but restrictive, convention to good effect.

We can sometimes simplify an expression by giving a meaning to extreme cases of notation. Consider the display

$$
\beta_{ij} = \begin{cases} 0, & i > j, \\[2mm] \dfrac{1}{u_j}, & i = j, \\[3mm] \dfrac{1}{u_j} \displaystyle\prod_{r=i}^{j-1} \left(\dfrac{-c_r}{u_r} \right), & i < j. \end{cases}
$$

There are really only two cases: $i > j$ and $i \leq j$. This structure is reflected and the display made more compact if we define the empty product to be 1, and write

$$\beta_{ij} = \begin{cases} 0, & \text{if } i > j, \\ \dfrac{1}{u_j} \displaystyle\prod_{r=i}^{j-1} \left(\dfrac{-c_r}{u_r} \right), & \text{if } i \leq j. \end{cases}$$

(Here, I have put "if" before each condition, which is optional in this type of display.) Incidentally, note that in a matrix product the order of evaluation needs to be specified: $\prod_{i=1}^{n} A_i$ could mean $A_1 A_2 \ldots A_n$ or $A_n A_{n-1} \ldots A_1$.

Notation also plays a higher level role in affecting the way a method or proof is presented. For example, the $n \times n$ matrix multiplication $C = AB$ can be expressed in terms of scalars,

$$c_{ij} = \sum_{k=1}^{n} a_{ik} b_{kj}, \qquad 1 \leq i, j \leq n,$$

or at the matrix-vector level,

$$C = [\, Ab_1, \quad Ab_2, \quad \ldots, \quad Ab_n \,],$$

where $B = [b_1, b_2, \ldots, b_n]$ is a partition into columns. One of these two viewpoints may be superior, depending on the circumstances. A deeper example is provided by the fast Fourier transform (FFT). The discrete Fourier transform (DFT) is a product $y = F_n x$, where F_n is the unitary Vandermonde matrix with (r, s) element $\omega^{(r-1)(s-1)}$ $(1 \leq r, s \leq n)$, and $\omega = \exp(-2\pi i / n)$. The FFT is a way of forming this product in $O(n \log n)$ operations. It is traditionally expressed through equations such as the following (copied from a numerical methods textbook):

$$\sum_{j=0}^{n-1} e^{2\pi i j k / n} f_j = \sum_{j=0}^{n/2-1} e^{2\pi i k j / (n/2)} f_{2j} + \omega^k \sum_{j=0}^{n/2-1} e^{2\pi i k j / (n/2)} f_{2j+1}.$$

The language of matrix factorizations can be used to give a higher level description. If $n = 2m$, the matrix F_n can be factorized as

$$F_n \Pi_n = \begin{bmatrix} I_m & \Omega_m \\ I_m & -\Omega_m \end{bmatrix} \begin{bmatrix} F_m & 0 \\ 0 & F_m \end{bmatrix},$$

where Π_n is a permutation matrix and $\Omega_m = \mathrm{diag}(1, \omega, \ldots, \omega^{m-1})$. This factorization shows that an n-point DFT can be computed from two $n/2$-point transforms, and this reduction is the gist of the radix-2 FFT. The book *Computational Frameworks for the Fast Fourier Transform* by Van

Loan [214], from which this factorization is taken, shows how, by using matrix notation, the many variants of the FFT can be unified and made easier to understand.

An extended example of how notation can be improved is given by Gillman in the appendix titled "The Use of Symbols: A Case Study" of *Writing Mathematics Well* [81]. Gillman takes the proof of a theorem by Sierpinski (1933) and shows how simplifying the notation leads to a better proof. Knuth set his students the task of simplifying Gillman's version even further, and four solutions are given in [128, §21].

Mathematicians are always searching for better notation. Knuth [127] describes two notations that he and his students have been using for many years and that he thinks deserve widespread adoption. One is notation for the Stirling numbers. The other is the notation $[S]$, where S is any true-or-false statement. The definition is

$$[S] = \begin{cases} 1, & \text{if } S \text{ is true,} \\ 0, & \text{if } S \text{ is false.} \end{cases}$$

For example, the Kronecker delta can be expressed as $\delta_{ij} = [i = j]$. The square bracket notation will seem natural to those who program; indeed, Knuth adapted it from a similar notation in the 1962 book by Iverson that led to the programming language APL [111, p. 11]. The square bracket notation is used in the textbook *Concrete Mathematics* [88]; that book and Knuth's paper give a convincing demonstration of the usefulness of the notation.

Table 3.1 shows the date of first use in print of some standard symbols; some of them are not as old as you might expect. Not all these notations met with approval when they were introduced. In 1842 Augustus de Morgan complained (quoted by Cajori [36, p. 328]):

> Among the worst of barbarisms is that of introducing symbols which are quite new in mathematical, but perfectly understood in common, language. Writers have borrowed from the Germans the abbreviation $n!$ to signify $1.2.3.\ldots(n-1)n$, which gives their pages the appearance of expressing surprise and admiration that 2, 3, 4, etc., should be found in mathematical results.

3.6. Words versus Symbols

Mathematicians are supposed to like numbers and symbols, but I think many of us prefer words. If we have to choose between reading a paper dominated by symbols and one dominated by words then, all other things

Symbol	Name	Year of publication
∞	infinity	1655 (Wallis)
π	pi (3.14159...)	1706 (Jones)
e	e (2.71828...)	1736 (Euler)
i	imaginary unit ($\sqrt{-1}$)	1794 (Euler)
$\equiv$	congruence	1801 (Gauss)
$n!$	factorial	1808 (Kramp)
$\sum$	summation	1820 (Fourier)
$\binom{n}{k}$	binomial coefficient	1826 (von Ettinghausen)
$\prod$	product	1829 (Jacobi)
δ_{ij}	Kronecker delta	1868 (Kronecker)
$|z|$	absolute value	1876 (Weierstrass)
$O(f(n))$	big oh	1894 (Bachmann)

Table 3.1. First use in print of some symbols. Sources: [36, 88, 127].

being equal, most of us would choose the wordy paper, because we would
expect it to be easier to understand. One of the decisions constantly facing
the mathematical writer is how to express ideas: in symbols, in words, or
both. I suggest some guidelines.

- Use symbols if the idea would be too cumbersome to express in words,
  or if it is important to make a precise mathematical statement.

- Use words as long as they do not take up much more space than the
  corresponding symbols.

- Explain in words what the symbols mean if you think the reader might
  have difficulty grasping the meaning or essential feature.

Here are some examples.

(1) Define $C \in \mathbb{R}^{n \times n}$ by the property that vec(C) is the eigenvector corresponding to the smallest eigenvalue in magnitude of
A, where the vec operator stacks the columns of a matrix into
one long vector.

To make this definition using equations takes much more space, and is
not worthwhile unless the notation that needs to be introduced (in this
case, a name for $\min\{ |\lambda| : \lambda$ is an eigenvalue of $A \}$) is used elsewhere. A

possible objection to the above wordy definition of vec is that it does not specify in which order the columns are stacked, but that can be overcome by appending "taking the columns in order from first to last."

(2) Since $|g'(0)| > 1$, zero is a repelling fixed point, so x_k does not tend to zero as $k \to \infty$.

An alternative is

Since $|g'(0)| > 1$, 0 is a repelling fixed point, so $x_k \not\to 0$ as $k \to \infty$.

This sentence is only slightly shorter than the original and is harder to read—the symbols are beginning to intrude on the grammatical structure of the sentence.

(3) If $B \in \mathbb{R}^{n \times n}$ has a unique eigenvalue λ of largest modulus then $B^k \approx \lambda^k xy^T$, where $Bx = \lambda x$ and $y^T B = \lambda y^T$ with $y^T x = 1$.

The alternative of "where x and y are a right and left eigenvector corresponding to λ, respectively, and $y^T x = 1$" is cumbersome.

(4) Under these conditions the perturbed least squares solution $x + \Delta x$ can be shown to satisfy

$$\frac{\|\Delta x\|}{\|x\|} \leq \epsilon \kappa_2(A)\left(1 + \frac{\|b\|}{\|A\|\|x\|}\right) + \epsilon\,\kappa_2(A)^2 \frac{\|r\|}{\|A\|\|x\|} + O(\epsilon^2).$$

Thus the sensitivity of x is measured by $\kappa_2(A)$ if the residual r is zero or small, and otherwise by $\kappa_2(A)^2$.

Here, we have a complicated bound that demands an explanation in words, lest the reader overlook the significant role played by the residual r.

(5) If $y_1, y_2, \ldots, y_n$ are all $\neq 1$ then $g(y_1, y_2, \ldots, y_n) > 0$.

In the first sentence "all $\neq 1$" is a clumsy juxtaposition of word and equation and most writers would express the statement differently. Possibilities include

If $y_i \neq 1$ for $i = 1, 2, \ldots, n$, then $g(y_1, y_2, \ldots, y_n) > 0$.
If none of the y_i $(i = 1, 2, \ldots, n)$ equals 1, then $g(y_1, y_2, \ldots, y_n) > 0$.

If the condition were "$\neq 0$" instead of "$\neq 1$", then it could simply be replaced by the word "nonzero". In cases such as this, the choice between words and symbols in the text (as opposed to in displayed equations) is a matter of taste; good taste is acquired by reading a lot of well-written mathematics.

The symbols $\forall$ and $\exists$ are widely used in handwritten notes and are an intrinsic part of the language in logic. But generally, in equations that are in-line, they are better replaced by the equivalent words "for all" and "there exists". In displayed equations either the symbol or the phrase is acceptable, though I usually prefer the phrase. Compare

$$\sigma\big(G(t)\big) = \exp\big(t\sigma(A)\big) \quad \text{for all } t \geq 0$$

with

$$\sigma\big(G(t)\big) = \exp\big(t\sigma(A)\big) \quad \forall t \geq 0.$$

Similar comments apply to the symbols $\Rightarrow$ and $\Longleftrightarrow$, though these symbols are more common in displayed formulas.

Of course, for some standard phrases that appear in displayed formulas, there is no equivalent symbol:

$$\text{minimize} \quad c^T x - \mu \sum_{i=1}^{n} \ln x_i \quad \text{subject to } Ax = b$$

$$\varliminf \tfrac{1}{n} \log D_j^n < 0 \quad \text{almost surely}$$

$$z^T y = \|z\|_D \|y\| = 1, \quad \text{where} \quad \|z\|_D = \max_{v \neq 0} \frac{|z^T v|}{\|v\|}.$$

3.7. Displaying Equations

An equation is displayed when it needs to be numbered, when it would be hard to read if placed in-line, or when it merits special attention, perhaps because it contains the first occurrence of an important variable. The following extract gives an illustration of what and what not to display.

Because $\delta(\overline{x}, \mu)$ is the smallest value of $\|\overline{X}z/\mu - e\|$ for all vectors y and z satisfying $A^T y + z = c$, we have

$$\delta(\overline{x}, \mu) \leq \|\tfrac{1}{\mu}\overline{X}z - e\|.$$

Using the relations $z = \mu X^{-1}s$ and $\overline{x}_i = 2x_i - x_i s_i$ gives

$$\frac{1}{\mu}\overline{X}z = \overline{X}X^{-1}s = (2X - XS)X^{-1}s = 2s - S^2 e.$$

Therefore, $\delta(\overline{x}, \mu) \leq \|2s - S^2 e - e\|$, which means that

$$\delta(\overline{x}, \mu)^2 \leq \sum_{i=1}^{n}(2s_i - s_i^2 - 1)^2 = \sum_{i=1}^{n}(s_i - 1)^4 \leq \left(\sum_{i=1}^{n}(s_i - 1)^2\right)^2 = \delta(x, \mu)^4.$$

The condition $\delta(x, \mu) < 1$ thus ensures that the Newton iterates $\overline{x}$ converge quadratically.

The second and third displayed equations are too complicated to put in-line. The first $\delta(\overline{x}, \mu)$ inequality is displayed because it is used in conjunction with the second display and it is helpful to the reader to display both these steps of the argument. The consequent inequality $\delta(\overline{x}, \mu) \leq \|2s - S^2 e - e\|$ fits nicely in-line, and since it is used immediately it is not necessary to display it.

When a displayed formula is too long to fit on one line it should be broken before a binary operation. Example:

$$|e_{m+1}| \leq |G^{m+1}e_0| + c_n u(1 + \theta_x)\{c(A)|(I - G)^D M^{-1}| + (m + 1)|(I - E)M^{-1}|\}(|M| + |N|)|x|.$$

The indentation on the second line should take the continuation expression past the beginning of the left operand of the binary operation at which the break occurred, though, as this example illustrates, this is not always possible for long expressions. A formula in the text should be broken after a relation symbol or binary operation symbol, not before.

3.8. Parallelism

Parallelism should be used, where appropriate, to aid readability and understanding. Consider this extract:

The Cayley transform is defined by $C = (A - \theta_1 I)^{-1}(A - \theta_2 I)$. If λ is an eigenvalue of A then

$$(\lambda - \theta_2)(\lambda - \theta_1)^{-1}$$

is an eigenvalue of C.

The factors in the eigenvalue expression are presented in the reverse order to the factors in the expression for C. This may confuse the reader, who might, at first, think there is an error. The two expressions should be ordered in the same way.

Parallelism works at many levels, from equations and sentences to theorem statements and section headings. It should be borne in mind throughout the writing process. If one theorem is very similar to another, the statements should reflect that—the wording should not be changed just for the sake of variety (see *elegant variation*, §4.16). However, it is perfectly acceptable to economize on words by saying, in Theorem 2 (say) "Under the conditions of Theorem 1."

For a more subtle example, consider the sentence

It is easy to see that $f(x, y) > 0$ for $x > y$.

In words, this sentence is read as "It is easy to see that $f(x, y)$ is greater than zero for x greater than y." The first $>$ translates to "is greater than" and the second to "greater than", so there is a lack of parallelism, which the reader may find disturbing. A simple cure is to rewrite the sentence:

It is easy to see that $f(x, y) > 0$ when $x > y$.
It is easy to see that if $x > y$ then $f(x, y) > 0$.

3.9. Dos and Don'ts of Mathematical Writing

Punctuating Expressions

Mathematical expressions are part of the sentence and so should be punctuated. In the following display, all the punctuation marks are necessary. (The second displayed equation might be better moved in-line.)

The three most commonly used matrix norms in numerical analysis are particular cases of the Hölder p-norm

$$\|A\|_p = \max_{x \neq 0} \frac{\|Ax\|_p}{\|x\|_p}, \qquad A \in \mathbb{R}^{m \times n},$$

where $p \geq 1$ and

$$\|x\|_p = \left(\sum_{i=1}^{n} |x_i|^p \right)^{1/p}.$$

Otiose Symbols

Do not use mathematical symbols unless they serve a purpose. In the sentence "A symmetric positive definite matrix A has real eigenvalues" there is no need to name the matrix unless the name is used in a following sentence. Similarly, in the sentence "This algorithm has $t = \log_2 n$ stages," the "$t =$" can be omitted unless t is defined in this sentence and used immediately. Watch out for unnecessary parentheses, as in the phrase "the matrix $(A - \lambda I)$ is singular".

Placement of Symbols

Avoid starting a sentence with a mathematical expression, particularly if a previous sentence ended with one, otherwise the reader may have difficulty parsing the sentence. For example, "A is an ill-conditioned matrix" (possible confusion with the word "A") can be changed to "The matrix A is ill-conditioned."

Separate mathematical symbols by punctuation marks or words, if possible, for the same reason.

Bad: If $x > 1$ $f(x) < 0$.
Fair: If $x > 1$, $f(x) < 0$.
Good: If $x > 1$ then $f(x) < 0$.

Bad: Since $p^{-1} + q^{-1} = 1$, $\| \cdot \|_p$ and $\| \cdot \|_q$ are dual norms.
Good: Since $p^{-1} + q^{-1} = 1$, the norms $\| \cdot \|_p$ and $\| \cdot \|_q$ are dual.

Bad: It suffices to show that $\|H\|_p = n^{1/p}$, $1 \le p \le 2$.
Good: It suffices to show that $\|H\|_p = n^{1/p}$ for $1 \le p \le 2$.
Good: It suffices to show that $\|H\|_p = n^{1/p}$ $(1 \le p \le 2)$.

Bad: For $n = r$ (2.2) holds with $\delta_r = 0$.
Good: For $n = r$, (2.2) holds with $\delta_r = 0$.
Good: For $n = r$, inequality (2.2) holds with $\delta_r = 0$.

"The" or "A"

In mathematical writing the use of the article "the" can be inappropriate when the object to which it refers is (potentially) not unique or does not exist. Rewording, or changing the article to "a", usually solves the problem.

Bad: Let the Schur decomposition of A be QTQ^*.
Good: Let a Schur decomposition of A be QTQ^*.

Good: Let A have the Schur decomposition QTQ^*.

Bad: Under what conditions does the iteration converge to the solution of $f(x) = 0$?

Good: Under what conditions does the iteration converge to a solution of $f(x) = 0$?

Notational Synonyms

Sometimes you have a choice of notational synonyms, one of which is preferable. In the following examples, the first of each pair is, to me, the more aesthetically pleasing or easier to read (a capital letter denotes a matrix).

$$\left(\sum_{i,j}(a_{ij} - b_{ij})^2\right)^{1/2}, \qquad \sqrt{\sum_{i,j}(a_{i,j} - b_{i,j})^2},$$

$$\exp\big(2\pi i(x^2 + y^2)^{-1/2}\big), \qquad e^{\frac{2\pi i}{\sqrt{x^2+y^2}}},$$

$$(1 - n\epsilon)^{-1}|L||U|, \qquad \frac{|L||U|}{1 - n\epsilon},$$

$$X_{k+1} = \tfrac{1}{2}X_k(3I - X_k^2), \qquad X_{k+1} = \frac{X_k}{2}[3I - X_k^2],$$

$$\min\{\,\epsilon : |b - Ay| \le \epsilon|A||y|\,\}, \qquad \min\{\,\epsilon \mid |b - Ay| \le \epsilon|A||y|\,\},$$

$$\min\{\,\|A - UBP\| : U^TU = I,\ P \text{ a permutation}\,\},$$

$$\min_{\substack{U^TU=I \\ P \text{ a permutation}}} \|A - UBP\|.$$

In the next two examples, the first form is preferable because it saves space without a loss of readability.

$$x = [\,x_1,\ \ x_2,\ \ \ldots,\ \ x_n\,]^T, \qquad x = \begin{bmatrix} x_1 \\ x_2 \\ \vdots \\ x_n \end{bmatrix},$$

$$\Lambda = \operatorname{diag}(\lambda_i), \qquad \Lambda = \begin{bmatrix} \lambda_1 & & & \\ & \lambda_2 & & \\ & & \ddots & \\ & & & \lambda_n \end{bmatrix}.$$

Of course, the diag($\cdot$) notation should be defined if it is not regarded as standard.

Miscellaneous

When working with complex numbers it is best not to use "i" as a counting index, to avoid confusion with the imaginary unit.

Note the difference between the Greek letter epsilon, ϵ, and the "belongs to" symbol $\in$, as in $\|x\| \leq \epsilon$ and $x \in \mathbb{R}^n$. Note the distinction between the Greek letter π and the product symbol $\prod$.

By convention, standard mathematical functions such as sin, cos, arctan, max, gcd, trace and lim are set in roman type, as are multiple-letter variable names. It is a common mistake to set these in italic type, which is ambiguous. For example, is $tanx$ the product of four scalars or the tangent of x?

In bracketing multi-layered expressions you have a choice of brackets for the layers and a choice of sizes, for example $\{[(\{[($, this ordering being the one recommended by *The Chicago Manual of Style* [212]. Most authors try to avoid mixing different brackets in the same expression, as it leads to a rather muddled appearance.

Write "the kth term", not "the kth term", "the k'th term" or "the k-th term." (It is interesting to note that nth is a genuine word that can be found in most dictionaries.)

A slashed exponent, as in $y^{1/2}$, is generally preferable to a stacked one, as in $y^{\frac{1}{2}}$.

Avoid tall in-line expressions, such as $\begin{bmatrix} g_1 \\ g_2 \end{bmatrix}$, which can disrupt the line spacing.

There are two different kinds of ellipsis: vertically centred ($\cdots$) and "ground level" or "baseline" ($\ldots$). Generally, the former is used between operators such as $+$, $=$, and $\leq$, and the latter is used between a list of symbols. Examples:

$$x_1 + x_2 + \cdots + x_n, \quad \sigma_1 \geq \sigma_2 \geq \cdots \geq \sigma_n, \quad \lambda_1, \lambda_2, \ldots, \lambda_n, \quad A_1 A_2 \ldots A_n.$$

An operator or comma should be symmetrically placed around the ellipsis; thus $x_1 + x_2 + \cdots x_n$ and $\lambda_1, \lambda_2, \ldots \lambda_n$ are incorrect.

When an ellipsis falls at the end of a sentence there is the question of how the full stop (or period) is treated. Recommendations vary. *The Chicago Manual of Style* suggests typing the full stop before the three ellipsis points (so that there is no space between the first of the four dots and the preceding character). When the ellipsis is part of a mathematical formula it seems natural to put it before the full stop, but the two possibilities may be visually indistinguishable, as in the sentence

The Mandelbrot set is defined in terms of the iteration $z_{k+1} = z_k^2 + c$, $k = 0, 1, 2, \ldots$.

A vertically centred dot is useful for denoting multiplication in expressions where terms need to be separated for clarity:

$$16046641 = 13 \cdot 37 \cdot 73 \cdot 457,$$

$$\mathrm{cond}(A, x) = \frac{\|\, |I - A^+ A| \cdot |A^T| \cdot |A^{+^T} x|\, \|}{\|x\|}.$$

Care is needed to avoid ambiguity in slashed fractions. For example, the expression $-(b-a)^3/12 f''(\eta)$ is better written as $-\big((b-a)^3/12\big) f''(\eta)$ or $-f''(\eta)(b-a)^3/12$.

Chapter 4
English Usage

There's almost no more beautiful sight than a simple declarative sentence.
— WILLIAM ZINSSER, *Writing with a Word Processor* (1983)

Quite aside from format and style,
mathematical writing is supposed to say something.
Put another way: the number of ideas divided by the
number of pages is supposed to be positive.
— J. L. KELLEY, *Writing Mathematics* (1991)

Let us not deceive ourselves.
"There is no God but Allah" is a more gripping sentence than
Mohammed (also Mahomet, Muhammad; 570?–632) asserted a doctrine of
unqualified monotheism (suras 8, 22, 33–37, 89, 91, Koran).
— MARY-CLAIRE VAN LEUNEN, *A Handbook for Scholars* (1992)

In this chapter I discuss aspects of English usage that are particularly relevant to mathematical writing. Three things need to be realized when reading this chapter. First, on many matters of English usage rules have exceptions, and, moreover, not all authorities agree on the rules. I have consulted several usage guides (those described in Chapter 2) and have tried to give a view that reflects usage in writing today. Second, about half the topics discussed here are not peculiar to the English language, but simply reflect common sense in writing. Third, many of the points mentioned are not vitally important when taken in isolation. But, as van Leunen explains (quoted in [128, p. 97]),

> Tone is important, and tone consists entirely of making these tiny, tiny choices. If you make enough of them wrong ... then you won't get your maximum readership. The reader who has to read the stuff will go on reading it, but with less attention, less commitment than you want. And the reader who doesn't have to read will stop.

4.1. A or An?

Whether *a* or *an* should precede a noun depends on how the first syllable is pronounced: *a* is used if the first syllable begins with a consonant sound and *an* if it begins with a vowel sound. This question most frequently occurs with acronyms, abbreviations and proper nouns. An easy example is "an NP-complete problem". In the world of mathematical software, given that the usual pronunciation of LAPACK is l-a-pack, and that of NAG is nag, we would write "an LAPACK routine" but "a NAG library routine".

4.2. Abbreviations

One school of thought says that the use of the abbreviations *e.g.* and *i.e.* is bad style, and that *for example* and *that is* make for a smoother flowing sentence. In any case, *i.e.* and *that is* should usually be followed by a comma, and all four forms should be preceded by a comma. When using the abbreviations you should type *e.g.* and *i.e.*, not *eg.* or *ie.*, since the abbreviations represent two words (the Latin *exempli gratia* and *id est*). Note that in the following sentence *i.e.* should be deleted: "The most expensive method, i.e. Newton's method, converges quadratically."

The less frequently used *cf.* has only one full stop, because it is an abbreviation of a single word: the Latin *confer*, meaning compare. Often, cf. is used incorrectly in the sense of "see", as in "cf. [6] for a discussion". The abbreviation *et al.* is short for *et alia*, so it needs only one full stop.

The abbreviation *N.B.* (of the Latin *nota bene*) is not often used in technical writing, probably because it has to appear at the beginning of a sentence and is somewhat inelegant. You can usually find a better way to give the desired emphasis.

The abbreviation *iff*, although handy in notes, is usually spelled out as *if and only if.*

The normal practice when introducing a nonstandard abbreviation is to spell out the word or phrase in full on its first occurrence and place the abbreviation immediately after it in parentheses. Thereafter the abbreviation is used. Example:

> Gaussian elimination (GE) is a method for solving a system of n linear equations in n unknowns. GE has a long history; a variant of it for solving systems of three equations in three unknowns appears in the classic Chinese work "Chiu-Chang Suan-Shu", written around 250 B.C.

4.3. Absolute Words

Certain adjectives have an absolute meaning and cannot be qualified by words such as less, quite, rather and very. For example, it is wrong to write *most unique* (replace by *unique*, or perhaps *most unusual*), *absolutely essential*, *more ideal*, or *quite impossible*. However, *essentially unique* is an acceptable term in mathematical writing: it means unique up to some known transformations. Many other words are frequently used with an absolute sense but can be modified (although for some words this usage is open to criticism); example phrases are "convergence is *almost certain*", "a *very complete* survey", "the *most obvious* advantage".

4.4. Active versus Passive

Prefer the active to the passive voice (prefer "X did Y" to "Y was done by X"). The active voice adds life and movement to writing, whereas too much of the passive voice weakens the communication between writer and reader.

> Passive: The answer was provided to sixteen decimal places by Gaussian elimination.

> Active: Gaussian elimination gave the answer to sixteen decimal places.

> Passive: The failure of Newton's method to converge is attributed to the fact that the Jacobian is singular at the solution.

Active: Newton's method fails to converge because the Jacobian is singular at the solution.

Passive: A numerical example is now given to illustrate the above result.

Active: We give a numerical example to illustrate this result, *or* The following numerical example illustrates this result.

The second example in the following trio illustrates a further degree of abstraction in which a verb is replaced by an abstract noun modified by another verb.

Passive: The optimality of y was verified by checking that the Hessian matrix was positive definite.

Passive and indirect: Verification of the optimality of y was achieved by checking that the Hessian matrix was positive definite.

Active: We verified the optimality of y by checking that the Hessian matrix was positive definite.

Other "was" phrases that can often be removed by rewriting are "was performed", "was experienced", "was carried out", "was conducted" and "was accomplished".

The passive voice has an important role to play, however. It adds variety, may be needed to put emphasis on a certain part of a sentence, and may be the only choice if specific information required for an active variant is unknown or inappropriate to mention. Examples where the passive voice allows the desired emphasis are "An ingenious proof of this conjecture was constructed by C. L. Ever," and (from the writings of Halmos [182, p. 96]) "The subjects just given honorable mention, as well as the three actually discussed in detail, have been receiving serious research attention in the course of the last twenty years."

4.5. Adjective and Adverb Abuse

Use an adjective only if it earns its place. The adjectives *very, rather, quite, nice* and *interesting* should be used with caution in technical writing, as they are imprecise. For example, in "The proof is very easy" and "This inequality is quite important" the adjectives are best omitted. Examples of acceptable usage are "These results are very similar to those of Smith" and "This bound can be very weak". *Interesting* is an overworked adjective that can often be avoided. For example, in the sentence "It is interesting

---

The Ten Commandments of Good Writing

1. Each pronoun should agree with their antecedent.

2. Just between you and I, case is important.

3. A preposition is a poor word to end a sentence with.

4. Verbs has to agree with their subject.

5. Don't use no double negatives.

6. Remember to never split an infinitive.

7. When dangling, don't use participles.

8. Join clauses good, like a conjunction should.

9. Don't write a run-on sentence it is difficult when you got to punctuate it so it makes sense when the reader reads what you wrote.

10. About sentence fragments.

Reprinted, with permission, from *How to Write and Publish a Scientific Paper* [51]

---

to consider how this result might be generalized," *instructive* is probably the intended word.

Try to avoid using nouns as adjectives. "The method for iteration parameter estimation" can be expressed more clearly as "The method for estimating iteration parameters." While proper nouns are often used as adjectives in speech ("this sequence is Cauchy", "that matrix is Toeplitz"), such usage in formal writing is best avoided (say "this is a Cauchy sequence", "that is a Toeplitz matrix"). Similarly, say "Euler's method is unstable" instead of "Euler is unstable".

An *adverb* that is overworked in mathematical writing is *essentially*. Dictionaries define it to mean necessarily or fundamentally, but it is often used with a vague sense meaning "almost, but not quite". Before using the word, consider whether you can be more precise.

> Bad: Beltrami (1873) essentially derived the singular value decomposition.
>
> Good: Beltrami (1873) derived the singular value decomposition for square, nonsingular matrices.

A valid use of *essentially* is in the expression "essentially the same as", which by convention in scientific writing means "the same, except for minor details".

4.6. -al and -age

Certain words that can be extended with an -al or -age ending are often misused in the extended form. The suffix tends to give a more abstract meaning, which makes it more difficult to use the word correctly. For example, *usage* means a manner of using, so correct *usage* is illustrated by "in the original usage the conjugate gradient method was not preconditioned" and "the use of Euler's method is not recommended for stiff differential equations."

An example where an -al ending is used incorrectly is "the most pragmatical opinion is the one expressed by the term's inventor," in which the third word should be "pragmatic".

4.7. Ambiguous "This" and "It"

A requirement of good writing is to make clear to the reader, at all times, what is the entity under discussion. *This* phrases such as "This is a consequence of Theorem 2" should be used with caution as they can force the reader to backtrack to find what "this" refers to. Often it helps to insert an appropriate noun after *this*. *It* can also be ambiguous: in the sentence "Condition 3 is not satisfied for the steepest descent method, which is why we do not consider it further" we cannot tell whether it is the condition or the method that is not being pursued.

4.8. British versus American Spelling

In my opinion (as a Briton) it makes little difference whether you use British or American (US) spellings, as long as you are consistent within a given piece of writing. For some journals, copy editors will convert a manuscript to one or the other form of spelling. Major dictionaries give both spellings. I find it natural to use the spelling of the country in which I am working at the time! See §5.6 for some examples of the different spellings.

4.9. Capitalization

Words that are derived from a person's name inherit the capitalization. Thus: Gaussian elimination, Hamiltonian system, Hermitian matrix, Jaco-

Correct/Intended	Misspelling
analogous	analagous
dependent	dependant*
discrete	discreet*
idiosyncrasy	idiosyncracy
in practice	in practise
led (past tense of lead)	lead*
lose	loose* (very common)
phenomenon	phenomena (plural of phenomenon)
preceding	preceeding
referring	refering
separate	seperate
supersede	supercede
zeros	zeroes[a]

[a]Some, but not all, dictionaries give the -oes ending as an alternative spelling for the plural noun.

Figure 4.1. Common errors.

bian matrix, Lagrangian function, Euler's method, and so on. The incorrect forms "hermitian" (capitalization) and "Lagrangean" (spelling) are sometimes seen. The Lax Equivalence Theorem is quite different from a lax equivalence theorem!

There does not seem to be a standard rule for when to capitalize the word following a colon. Knuth [128, p. 11] suggests the useful convention of capitalizing when the phrase following the colon is a full sentence.

4.10. Common Misspellings or Confusions

The errors shown in Figure 4.1 seem to be common in mathematical writing. The misspellings marked with an asterisk are genuine words, but have a different meaning from the word in the left column.

According to McIlroy [146], on most days at Bell Laboratories someone misspells the word *accommodate*, in one of seven incorrect ways.

4.11. Compound Words

As Turabian [207, p. 44] notes, the trend is not to hyphenate compound words beginning with prefixes such as multi, pre, post, non, pseudo and semi. In mathematical writing it is common to write *nonsingular, semidef-*

inite (but *semi-infinite* to avoid a double i) and *pseudorandom.* However, a hyphen is retained before a proper noun, as in *non-Euclidean.* In deciding whether to hyphenate or to combine two words as one, it is worth bearing in mind that the hyphenated form tends to be easier to read because the prefix can be seen at a glance. Readers whose first language is not English may appreciate the hyphenated form.

4.12. Consistency

It is important to be consistent. Errors of consistency can be very hard to detect, but can be puzzling to the reader when they are spotted. Don't refer to ker(A) as the null-space, or null(A) as the kernel—stick to matching synonyms. If you use the term "Cholesky factorization", don't say the "Cholesky decomposition" in the same work. If words have alternative spellings, stick to one: don't use both orthogonalise and orthogonalize, and if you use orthogonalize also use, for example, optimize, not optimise. But note that not all -ise words can be spelt with -ize; examples are listed in §5.6.

4.13. Contractions

Contractions such as *it's, let's, can't* and *don't* are not used in formal works, but are acceptable in popular articles if used sparingly. Note the distinction between the contraction *it's* (short for *it is*) and the possessive *its*: "It's raining", "*A* is singular if its determinant is zero". An apostrophe denotes possession for nouns (*the proof's length*) but not for pronouns (*this book is yours*).

4.14. Dangling Participle

What is wrong with the following sentence?

Substituting (3) into (7), the integral becomes $\pi^2/9$.

This sentence suggests that the integral makes the substitution. The error is that the intended subject ("we") of the participle *substituting* is not present in the sentence. Rewritten and unambiguous versions are

Substituting (3) into (7), we find that the integral is $\pi^2/9$.
When (3) is substituted into (7), the integral becomes $\pi^2/9$.

A similar example that is less obviously wrong is

When deriving parallel algorithms, the model of computation must be considered carefully.

Here is another example of a different type:

Being stiff, the Runge–Kutta routine required a large amount of CPU time to solve the differential equation.

Here, the problem is that the noun immediately following "being" is not the one to which this participle refers. There are several ways to rewrite the sentence. One that preserves the emphasis is

Because the differential equation is stiff, the Runge–Kutta routine required a large amount of CPU time to solve it.

Certain participial constructions are idiomatic and hence are regarded as acceptable:

Assuming $G(x^*)$ is positive definite, x^* is a minimum point for F.

Considering the large dimension of the problem, convergence was obtained in remarkably few iterations.

4.15. Distinctions

Affect, Effect. *Affect* is a verb meaning to produce a change. *Effect* is a noun meaning the result of a change. Examples: "Multiple roots affect the convergence rate of Newton's method," "One effect of multiple roots is to reduce the convergence rate of Newton's method to linear." Effect is also a verb meaning to bring about (as in "to effect a change"), but in this form it is rarely needed in mathematical writing.

Alternate, Alternative. *Alternate* implies changing repeatedly from one thing to another. An *alternative* is one of several options. Compare "While writing his thesis the student alternated between elation and misery," with "The first attempt to prove the theorem failed, so an alternative method of proof was tried."

Compare with, Compare to. *Compare with* analyses similarities and differences between two things, whereas *compare to* states a resemblance between them. Examples: "We now compare Method A with Method B," "Shakespeare compared the world to a stage," "Shall I compare thee to a summer's day?" As Bryson [29] explains, "Unless you are writing poetry or love letters, *compare with* is usually the expression you want."

Comprise, Compose, Constitute. *Compose* means to make up, *comprise* means to consist of. "Comprised of" is always incorrect. Thus, "the course is composed of three topics" or "the course comprises three topics", but not "the course is comprised of three topics". *Constitute* is a transitive verb used in the reverse sense: "these three topics constitute the course".

Fewer, Less. *Less* refers to quantity, amount or size, *fewer* to number. Thus "the zeros of $f(x)$ are less than those of $g(x)$" means that if x is a zero of f and y a zero of g then $x < y$, whereas "the zeros of $f(x)$ are fewer than those of $g(x)$" means that g has more zeros than f. Bryson [29] states the rule of thumb that *less* should be used with singular nouns and *fewer* with plural nouns: less research, less computation, fewer graduates, fewer papers.

Practice, Practise. In British English, *practice* is the noun and *practise* the verb (as with advice and advise). Thus "in practice", "practice session", "practise the technique", "practised speaker". But in American English both verb and noun are spelt *practice*.

Which, That. A "wicked which"[1] is an instance of the word *which* that should be *that* (example: replace the word before *should*, earlier in this sentence, by *which*). The rule is that *that* defines and restricts, whereas *which* informs and does not restrict.[2] (Mathematicians should be good at spotting definitions.) Note the difference between the following two examples.

"Consider the Pei matrix, which is positive definite." We are being told additional information about the Pei matrix—that it is positive definite.

"Consider the Pei matrix that is positive definite." Now we are being asked to focus on a particular Pei matrix from among several: the one that is positive definite.

A useful guide is that which-clauses are surrounded by commas, or preceded by a comma if at the end of a sentence. If you're not sure whether to use *which* or *that*, see whether your sentence looks right with commas around the relevant clause. Sometimes it pays to introduce a wicked which to avoid ugly repetition, as has been done in the sentence "This approach is similar to that which we used in our earlier paper" (though "the one we used" is better). A rule of thumb discussed in [128, pp. 96–97] is to replace *which* by *that* whenever it sounds right to do so.

4.16. Elegant Variation

Elegant variation is defined by the Fowlers [64] as "substitution of one word for another for the sake of variety." It is a tempting way to avoid repeti-

---

[1] A term coined by Knuth [128].

[2] Some authorities permit *which* to be used in a defining clause (e.g., [87]), but, as Bryson [29] puts it, "the practice is on the whole better avoided."

tion, but is often unnecessary and can introduce ambiguity. Consider the sentence "The eigenvalue estimate from Gershgorin's theorem is a crude bound, but it is easy to compute." Does Gershgorin's theorem yield an estimate or a strict bound? We cannot tell from the sentence. In fact, the answer is that it can yield either, depending on how you interpret the theorem. A rewrite of the sentence (with knowledge of the theorem) produces "The eigenvalue inclusion regions provided by Gershgorin's theorem are crude, but easy to evaluate," where *inclusion regions* can be replaced by *estimates* or *bounds*, depending on the desired emphasis.

The opposite of elegant variation is when the same word is repeated in different forms or with different meanings. Here are two examples.

> The performance is impressive and gives the impression that the blocksize is nearly optimal. [impressive, impression]
> In the remainder of this chapter we examine the remainder in Euler's summation formula. [remainder, remainder]

Such echoing of words is distracting and is easily avoided by choosing a synonym for one of them.

4.17. False If

The if–then construct is a vital tool in expressing technical arguments, but it is sometimes used incorrectly. Consider the sentence

> If we wish to compare the solutions of $f - \lambda k(f) = 0$ and $f_n - \lambda k_n(f_n) = 0$, then Jones shows that for a wide class of nonlinear $k(f)$, $\|f - f_n\| \leq c(\lambda)\|k_n(f) - k(f)\|$.

Jones's demonstration is not dependent on us wishing to compare solutions, so the *if* is misleading: it falsely heralds a logical condition. False ifs can always be removed by rewriting:

> To compare the solutions of $f - \lambda k(f) = 0$ and $f_n - \lambda k_n(f_n) = 0$, we can use Jones's result that for a wide class of nonlinear $k(f)$, $\|f - f_n\| \leq c(\lambda)\|k_n(f) - k(f)\|$.

A more confusing example is

> If we assume that rational fractions behave like almost all real numbers, a theorem of Khintchine states that the sum of the first k partial quotients will be approximately $k \log_2 k$.

The "if" appears to be a false one, because the statement of Khintchine's theorem must be independent of what the writer assumes. In fact, with knowledge of the theorem, it can be seen that the error is in the word "states", which should be "implies".

The next example is an unnecessary if, rather than a false if.

> We show that if $\widehat{x}$ is the computed solution to $Lx = b$ then $(L + \Delta L)\widehat{x} = b$, where $\|\Delta L\| \leq \alpha(n)\epsilon\|L\|$.

This type of construction is acceptable if used sparingly. I prefer

> We show that the computed solution $\widehat{x}$ to $Lx = b$ satisfies $(L + \Delta L)\widehat{x} = b$, where $\|\Delta L\| \leq \alpha(n)\epsilon\|L\|$.

4.18. Hyphenation

Compound words involving *ill* and *well* occur frequently in mathematical writing and opinions differ about their hyphenation. *The Chicago Manual of Style* [212] recommends hyphenation when a compound with *ill, well, better, best, little, lesser*, etc., appears before a noun, unless the compound is itself modified. The purpose of this hyphenation rule is to avoid ambiguity. Examples:

> This is an ill-posed problem *but* This problem is ill posed.
> The well-known theorem *but* The theorem is well known.
> An ill-conditioned function *but* A very ill conditioned function.
> The second-order term has a constant 2 *but* This term is of second order.

The second example is widely accepted, but many writers always hyphenate terms such as ill-conditioned, and it is hard to argue against this practice. There are some common phrases that some writers hyphenate and others do not. An example is *floating point arithmetic* (*floating-point arithmetic*).

Notice the hyphen in the title of Halmos's best seller *Finite-Dimensional Vector Spaces* [92]. Halmos explains in [98, p. 146] that in the original 1942 edition the hyphen was omitted, but it was added for the 1958 edition.

Hyphenation conventions can change over time, as certain terms and compound words gain wide acceptance (see *compound words*, §4.11).

4.19. Linking Words

If a piece of writing is to read well there must be no abrupt changes in mood or direction from sentence to sentence within a paragraph (unless such changes are used deliberately for effect). One way to achieve a smooth flow is to use linking words. Notice how the following paragraph is improved by adding "In particular" to the start of the second sentence and "Furthermore" to the start of the third.

> Once we move from a convex program to a general nonlinear program, matters become far more complicated. Certain topological assumptions are required to avoid pathological cases. The results apply only in a neighbourhood of a constrained minimizer, and involve convergence of subsequences of global minimizers of the barrier function.

Of course, a sequence of sentences of the form "adverb, fact" quickly becomes tiresome, so linking words should be not be overused.

Here is a list of link words and phrases, arranged according to sense. For examples of use see §5.5.

combinations. also, and, as well as, besides, both, furthermore, in addition to, likewise, moreover, similarly.

implications or explanations. as, because, conversely, due to, for example, given, in other words, in particular, in view of, it follows that, otherwise, owing to, since, specifically, that is, unlike.

modifications and restrictions. although, alternatively, but, despite, except, however, in contrast, in spite of, nevertheless, of course, on the contrary, on the other hand, though, unfortunately, yet.

emphasis. actually, certainly, clearly, in fact, indeed, obviously, surely.

consequences. accordingly, consequently, hence, therefore, thus.

4.20. Numbers

Small integers should be spelled out when used as adjectives ("The three lemmas"), but not when used as names or numbers ("The median age is 43" or "This follows from Theorem 3"). The number 1 is a special case, for often "one" or "unity" reads equally well or better: "z has modulus one".

4.21. Omit These Words?

Here are some words and phrases whose omission often improves a sentence:

actually, very, really, currently, in fact, thing, without doubt.

4.22. Paragraphs

A standard device for making text more appetizing is to break it into small paragraphs, as is done in newspapers. The short paragraph principle is worth following in technical writing, where the complexity of the ideas makes it more important than usual for the reader to be given frequent rests. Furthermore, long paragraphs tend to give a page a heavy image that can be a visual deterrent to the potential reader. A mix of different lengths is best. Ideally, each paragraph contains a main idea or thought that separates it from its neighbours. A long paragraph that is hard to break may be indicative of convoluted thinking.

The best writers occasionally slip in one-sentence paragraphs.

4.23. Punctuation

Much can be said about punctuation, and for thorough treatments of the topic I refer the reader to the references mentioned in Chapter 2. It is worth keeping in mind Carey's explanation [39] that "the main function of punctuation is *to make perfectly clear the construction* of the written words." A few common mistakes and difficulties deserve mention here.

- In "This result is well known, see [9]" the comma should be a semi-colon, which conveys a slightly longer pause. Even better is to say "This result is well known [9]." A common mistake is the let-comma-then construction: "Let x^* be a local maximum of $F(x)$, then a Taylor series expansion gives ...." The comma should be a full stop.

  Similarly, the comma should be a semicolon, or even a full stop, in this sentence: "This bound has the disadvantage that it uses a norm of X, moreover the multiplicative constant can be large when X is not a normal matrix."

- In this sentence the semicolon should be a comma: "The secant method can also be used; its lack of need for derivatives being an advantage." A rough guide is to use a semicolon only where you could also use a full stop.

- In the sentence "If we use iterative refinement solutions are computed to higher accuracy," a comma is needed after "refinement", otherwise the reader may take "iterative refinement" as modifying "solutions". Another example where a comma is needed to avoid ambiguity is the sentence "However, the singularity can be removed by a change of variable."

- In sentences such as

> Fortran 77 contains the floating point data types real, complex and double precision.
>
> The output can be rotated, stretched, reduced or magnified.

we have a choice of whether or not to place a comma (called a serial comma) before the *and* or *or* that precedes the final element of the list. In some sentences a serial comma is needed to avoid ambiguity, as in the sentence from page 6 "It is used to check spelling, shades of meaning, and usage", where the absence of the comma makes "shades" modify usage. Opinion differs on whether a serial comma should always be used. It is a matter of personal preference. The house styles of some publishers require serial commas.

If a list contains commas within the items, ambiguity can be avoided by using a semicolon as the list separator. Example:

> The test collection includes matrices with known inverses or known eigenvalues; ill-conditioned or rank deficient matrices; and symmetric, positive definite, orthogonal, defective, involutary, and totally positive matrices.

- The exclamation mark should be used with extreme caution in technical writing. If you are tempted to exclaim, read "!" as "shriek"; nine times out of ten you will decide a full stop is adequate. An example of correct usage is, from [160, p. 42], "When A is tridiagonal the computation of $A^{-1}u$ costs little more than the computation of Au!" The exclamation mark could be omitted, but then the reader might not realize that this is a surprising fact. Another example is, from [123, p. 42], "The chi-square table tells us, in fact, that V_2 *is much too low*: the observed values are so close to the expected values, we cannot consider the result to be random!"

4.24. Say It Better, Think It Gooder

The title of this section combines the titles of two papers by Piranian [169] and Halmos [96] that appeared in *The Mathematical Intelligencer* in 1982. As Gillman explains in [82], "George said good English is important. Paul said, what do you mean, good English is important?—good mathematics is more important. They are both right." While correct English usage is important, it must not be allowed to deflect you from the language-independent tasks of planning and organizing your writing.

4.25. Saying What You Mean

In technical writing you need to take great care to say what you mean. A hastily constructed sentence can have a meaning very different from the one intended. In a book review in *SIAM Review* [vol. 34, 1992, pp. 330–331] the reviewer quotes the statement "According to Theorem 1.1, a single trajectory $X(t,x)$ passes through almost every point in phase space ...." The book's author meant to say that for every point in phase space there is a unique trajectory that goes through it.

4.26. Sentence Opening

Try not to begin a sentence with *there is* or *there are*. These forms of the verb *be* make a weak start to a sentence (see the quote by Dixon on page 67). Sometimes these phrases can simply be deleted, as in the sentence "There are several methods that are applicable" ("Several methods are applicable"). Also worth avoiding, if possible, are "It is" openers, such as "It is clear that" and "It is interesting to note that". If you can find alternative wordings, your writing will be more fresh and lively.

4.27. Simplification

Each word or phrase in the left column below can (or, if marked with an asterisk, should) be replaced by the corresponding one in the right column. This is not a complete list (see [51, Appendix 4] or [11] for many more examples) but comprises periphrastic phrases that I have spotted in mathematical writing.

by means of	by
conduct an investigation	investigate
due to the fact that*	since/because
firstly	first
in a position to	can

in order to	to
in the case that*	if
in the course of	during
in the event that*	if
in the first place*	first
it is apparent that	apparently
it may happen that	there may/might
most unique*	unique
take into consideration	consider
that is to say	in other words, that is
to begin with*	to begin

4.28. Synonym Selection

English is one of the most synonym-rich languages, thanks to the words it has adopted from other languages, and each member of a set of synonyms can have a different tone and shade of meaning. A dictionary and a thesaurus are vital aids to choosing the right word (see Chapter 2). When you have a choice of words, use a short, concrete one in preference to a long or abstract one. Often, this means using an Anglo-Saxon word instead of one of French or Latin origin, as Table 4.1 illustrates.

Anglo-Saxon	French	Latin
ask	question	interrogate
rise	mount	ascend
good	marvellous	superior
show	establish	exhibit, demonstrate
need	requirement	necessity

Table 4.1. The origins of some synonyms.

Don't be afraid of using a long or unusual word if it is just the right one, or if it adds spice to your writing or conveys an interesting image. Parlett [160, p. 198] writes "How should X be chosen? The answer depends on whether or not there is any adscititious information about A besides H and C." The uncommon word *adscititious* (which is in the seventh edition (1982) but not in the eighth edition (1990) of the Concise Oxford Dictionary) means adopted from without. Acton [3, p. 145] writes "The second term ... is obstreperous." No other word quite so vividly describes the uncontrollable term to which he refers (with the possible exception of *recalcitrant*, another word used by Acton in the same book).

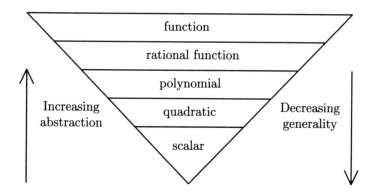

Figure 4.2. Word pyramid.

I chose the title "Otiose Symbols" on page 25 in preference to "Unnecessary Symbols" for two reasons. First, the former is shorter and more likely to catch the reader's eye. Second, *otiose* means "serving no practical purpose", so "otiose symbols" has a stronger meaning than "unnecessary symbols".

It helps to think of word pyramids showing levels of abstraction [210], such as the one in Figure 4.2. As Turk and Kirkman [209] point out, "Abstract words are less easy to decode because the reader has to 'scan' all the possibilities subconsciously before deciding on a specific meaning." Generally, it is best to use the least abstract, most specific word possible. This leads to more lively and precise prose. Here are some examples of chains of words in order of increasing specificity:

    result–theorem–inequality
    statistic–error–relative error
    optimum–minimum–global minimum
    random–normally distributed–normal (0,1)

4.29. Tense

One of the decisions the writer of a technical paper must make is what tense to use. There are no hard and fast rules, except to be consistent in the use of tense, but I offer some guidelines.

I prefer the present tense to the future tense for referring to later parts of the paper. Thus I write "This is proved in Theorem 3, below" rather than "This will be proved in Theorem 3", and "We discuss these matters in detail in section 5" rather than "We will discuss these matters in detail". Note that since the latter example does not contain a section reference it could be construed as referring to a future paper; this potential ambiguity is a danger of using the future tense.

The present tense can also be used for backward references, as in "The analysis of section 1 proves the following ...," though the past tense is more common: "We showed/saw in section 1 that ...." Tables and figures exist in the present, so the present tense should be used to refer to them: "Table 4 shows that", "Figure 1 displays". Similarly, in a citation where the reference is the object of the sentence, the present tense is the one to use: "Reference [4] contains the interesting result that ...."

To refer to work in earlier papers either the past or the present tense can be used: "Banach shows that" or "Banach showed that". I use the present tense unless I want to emphasize the historical aspects of the reference, as, for example, when summarizing a number of earlier results in a survey.

A summary or conclusions section should use the past tense when specifying actions: "We have shown that" or "We showed that" rather than "We show that". But if a simple statement of results is given, the present tense is appropriate: "Asymptotic expansions are useful for ...," "Our conclusion is that ...."

In summary, I recommend the rule "if in doubt use the present tense".

4.30. What to Call Yourself

All technical writers face the question of how to refer to themselves in their papers. For example, if you refer to a previous result of yours you may have to choose between

(1) I showed in [3] that
(2) We showed in [3] that
(3) The author showed in [3] that
(4) It was shown in [3] that

Sentence 4 should be avoided because it is in the passive voice, which is unnecessary in this case. Sentence 3 has a formal, stilted air and is again best avoided. Some authorities on technical writing advise against using "we" to refer to a single author, but are happy for it to be used when there are two or more authors. Nevertheless, in mathematical writing "we" is by far the most common choice of personal pronoun; I frequently use it myself (in papers, but not in this book). "I" creates a more frank, personal contact with the reader and is less commonly used than "we". Zinsser [225, p. 24,] suggests "If you aren't allowed to use 'I', at least think 'I' while you write,

or write the first draft in the first person and then take the 'I's out. It will
warm up your impersonal style."

"We" can be used in the sense of "the reader and I", as in the sentence

We saw earlier that $f_n \to 0$ as $n \to \infty$.

Whether you choose "I" or "we", you should not mix the two in a single
document, except, possibly, when using the "reader and I" form of "we".

Other personal pronouns to consider are "one" and "us". "One", as in
"one can show that ..." is often used, but is perhaps best avoided because
of its vague, impersonal nature. "Us" is useful in sentences such as

Let us now see how the results can be extended to non-smooth
functions.

where it means "the reader and I". It is generally quite easy to remove both
"one" and "us" by phrasing sentences in an alternative form. For example,
the sentence above can be rewritten as "How can the results be extended
to non-smooth functions? One approach is ...," or "To see how the results
can be extended to non-smooth functions, we ...."

4.31. Word Order

The first words of a sentence are usually regarded by the reader as the
most important. Therefore, the word order of sentences should be chosen
to reflect the desired emphasis. Compare the first sentence of this paragraph
with "The reader usually regards the first words of a sentence as the most
important," which places the emphasis on the reader rather than "the first
words of a sentence".

Reordering the words of a sentence can strengthen it and remove ambiguity. Here is an example of a misplaced *only*.

Bad: The limit point is only a stationary point when the regu-
       larity conditions are satisfied. (Might be taken to mean
       that the limit point is expected to be more than just a
       stationary point.)

Good: The limit point is a stationary point only when the reg-
       ularity conditions are satisfied.

Here is an example of an incorrect *either*.

Bad: Mathematically, either we have an integral equation of
       the first kind or one of the second kind.

Good: Mathematically, we have an integral equation of either
the first kind or the second kind. (Or simply delete *either*
from the bad example.)

Chapter 5
When English Is a Foreign Language

Two authors of the present book . . .
freely acknowledge that they are more fluent in German and French,
yet they enjoy communicating in English and are convinced that it is
more than mere historical coincidence . . . that has led English
to become the lingua franca *of modern science.*
— HANS F. EBEL, CLAUS BLIEFERT, and WILLIAM E. RUSSEY,
The Art of Scientific Writing (1987)

I ventured to write this book in English because
it will be more easily read in poor English
than in good German by 90% of my intended readers.
— HANS J. STETTER, *Analysis of Discretization Methods for*
Ordinary Differential Equations (1973)

If English is not your first language then writing mathematics is doubly
difficult, since you have to grapple with grammar and vocabulary as you
try to express your thoughts. In this chapter I offer a few basic suggestions
and give some references for further reading. The problems encountered by
a non-native writer of English depend very much on the writer's first language. I concentrate on general advice that is not specific to any particular
language.

5.1. Thinking in English

Try to think in English and construct sentences in English. If you compose
in your own language and then translate you are more likely to produce
prose that is not idiomatic (that is, does not agree with standard usage).
Here are some examples of non-idiomatic phrases or sentences:

> ▷ *Capable to do.* Although this is logically correct, conven-
> tion requires that we say "capable of doing".

> ▷ *We have the possibility to obtain an asymptotic series for
> the solution.* We do not normally say "possibility to". Bet-
> ter is *It is possible to obtain* ... (passive voice), or, shorter,
> *We can obtain* ....

> ▷ *This result was proved already in* [5]. *Already* should be
> deleted or replaced by *earlier* or *previously*. Alternative:
> *This result has already been proved in* [5].

> ▷ *The solution has been known since ten years.* This type of
> construction occurs in those European languages in which
> one word serves for both *for* and *since*. In this example,
> *since* should be replaced by *for*.

> ▷ *This approach permits to exploit the convexity of f.* The
> phrase should be *permits us to exploit* (active voice) or
> *permits exploitation of* (passive voice). Or, depending on
> the context, it may be acceptable to shorten the sentence
> to *This approach exploits the convexity of f*.

Part of the process of thinking in English is to write your research notes
in English from the start and to annotate the books and papers you study
in English.

5.2. Reading and Analysing Other Papers

Read as many well-written papers in your field as you can. Ask a friend or
colleague for recommendations. Analyse the following aspects.

Vocabulary. What kinds of words occur frequently in technical writing? Look up words that you don't know in a dictionary. For technical terms you may need to use a mathematical or scientific dictionary or encyclopedia. Note the range of applicability of particular words.

Synonyms. Notice different ways of saying the same thing and different ways of ordering sentences. Use what you learn to avoid monotony in your writing.

Collocations. These are groups of words that commonly appear together. For example, feeble and fragile are synonyms for weak, but weak has more meanings, and while we readily say "weak bound" we never say "feeble bound" or "fragile bound". As another example, we say "uniquely determined" or "uniquely specified" but not "uniquely fixed" or "uniquely decided". Build up a list of the collocations you find in mathematical writing.

Idioms. Idioms are expressions whose meanings cannot be deduced from the words alone, but are established by usage. Here are some examples of idioms that are sometimes found in technical writing and speaking, with the idiom in the left column and a definition in the right column.

By and large.	Taking everything into account.
End up.	Reach a state eventually.
In that.	In so far as.
It goes without saying.	Something so obvious that it needn't be said.
On the other hand.	From the other point of view.
On the whole.	In general, ignoring minor details.
Over and above.	In addition to.
Rule of thumb.	A rule based on experience and estimation rather than precise calculation.
Start from scratch.	To start from the beginning with no help.

Errors in the use of idioms tend to be very conspicuous. It is good advice to avoid idioms until you are sure how to use them correctly.

5.3. Articles

Some languages either do not have articles (words such as "the", "a" and "an") or use them in a different way than in English, so it is difficult for speakers of these languages to use articles correctly in English. The rules of article use are complicated. Swan [201] explains them and identifies two of the most important.

(1) Do not use *the* (with plural or uncountable nouns) to talk about things in general. Examples: "Mathematics is interesting" (not "The mathematics is interesting"); "Indefinite integrals do not always have closed form

LORETTE WILMOT LIBRARY
NAZARETH COLLEGE

solutions" (not "The indefinite integrals do not always have the closed form solutions").

(2) Do not use singular countable nouns without articles. Examples: "the derivative is", "a derivative is", but not "derivative is".

Mistakes in the use of articles are undesirable, but they do not usually obscure the meaning of a sentence.

5.4. Constructions

Certain constructions are common in mathematical writing. You may find it helpful to make a list for reference, beginning with the following entries. The left column contains constructions and the right column examples.

Let ... be	Let f be a continuous function.
If ... then	If $\alpha > -1$ then the integral exists.
Suppose (that) ... is/are	Suppose g is differentiable.
	Suppose that A and B have no eigenvalue in common.
We define ... to be	We define a problem to be stable if ...
It is easy to see/show that	It is easy to show that the error decays as t increases.
From ... we have	From (5.2) we have the inequality ...
By substituting ... into ... we obtain	By substituting (1.9) into (7.3) we obtain ...
A lower bound for	A lower bound for h can be obtained from ...

5.5. Connecting Words and Phrases

In this section I give examples of the use of words and phrases that connect statements. Most of the examples are followed by comments on the degree of emphasis; note, however, that the emphasis imparted sometimes depends on the context in which the word or phrase appears, so extrapolation from these examples should be done with care. Mastering these connectives, and the differences between them, is an important part of learning to write technical English. This section is loosely based on [59, pp. 191–194].

Combinations

Statement a: Direct methods are used to solve linear systems.
Statement b: Iterative methods are used to solve linear systems.

and   Direct methods *and* iterative methods are used to solve linear systems.
                 (No emphasis on either type of method.)

both   *Both* direct and iterative methods are used to solve linear systems.
                 (Similar to *and*.)

also   Direct methods, and also iterative methods, are used to solve linear systems.
                 Direct methods are used to solve linear systems, as also are iterative methods.
                 (Slight emphasis on direct methods.)

as well as Direct methods, as well as iterative methods, are used to solve linear systems.
                 (Similar to *also*.)

not only ... **but also** Linear systems can be solved *not only* by direct methods *but also* by iterative methods.
                 (Emphasizes that there is more than one possibility.)

apart from/in addition to *Apart from (in addition to)* direct methods, iterative methods are used to solve linear systems.
                 (Emphasizes that there is more than one type of method; slightly more emphasis than *also* but less than *not only ... but also*.)

moreover/furthermore The name of Gauss is attached to the most well-known method for solving linear systems, Gaussian elimination. *Moreover (furthermore)*, a popular iterative technique also takes his name: the Gauss–Seidel method.
                 (Stresses the statement after *moreover/furthermore*.)

Implications or Explanations

Statement a: The problem has a large condition number.
Statement b: The solution is sensitive to perturbations in the data.

as/because/since *As (because, since)* the problem has a large condition number, the solution is sensitive to perturbations in the data.
                 The solution is sensitive to perturbations in the data, *as (because, since)* the problem has a large condition number.

due to	The sensitivity of the solution to perturbations in the data is *due to* the ill condition of the problem. (More emphatic than *as*.)
in view of/owing to/on account of	*In view of* (*owing to, on account of*)[1] the ill condition of the problem, the solution is sensitive to perturbations in the data. (More emphatic than *as*.)
given	*Given* the ill condition of the problem, the solution is necessarily sensitive to perturbations in the data. (Inevitable result of the stated condition.)
it follows that	The problem has a large condition number. *It follows that* the solution is sensitive to perturbations in the data. (Puts more emphasis on the first statement than *as*.)
therefore	The problem has a large condition number and *therefore* the solution is sensitive to perturbations in the data. (Intermediate between *as* and *it follows that*.)

Modifications and Restrictions

Statement a: Runge–Kutta methods are widely used for solving non-stiff differential equations.

Statement b: For stiff differential equations, methods based on backward differentiation formulae (BDF) are preferred.

alternatively	If the differential equations are non-stiff, Runge–Kutta methods can be used; *alternatively*, if the differential equations are stiff, BDF methods are preferred.
although	*Although* Runge–Kutta methods are widely used for non-stiff differential equations, BDF methods are preferred when the differential equations are stiff.

---

[1] *Due to* would be incorrect here, though this usage is acceptable in spoken English. For more details see [198, p. 45] and [221].

(More emphasis on BDF methods than *alternatively.*)

though   Runge–Kutta methods are widely used for non-stiff differential equations, *though* BDF methods are preferred when the differential equations are stiff.

(*Though* is weaker than *although*, and it tends to be used inside a sentence rather than at the beginning. In this example *though* could be replaced by *although*, which would give greater emphasis to the BDF methods.)

but   If the differential equations are non-stiff, Runge–Kutta methods can be used, *but* if the differential equations are stiff, BDF methods are preferred.

(Similar to *though.*)

except   *Except* for stiff differential equations, for which BDF methods are preferred, Runge–Kutta methods are widely used.

(Clearly defined limitation or restriction.)

however/on the other hand   Runge–Kutta methods are widely used for solving non-stiff differential equations. *However* (*on the other hand*), for stiff differential equations BDF methods are preferred.

(Note that *however* and *on the other hand* are not always interchangeable. *On the other hand* is applicable only when there are two possibilities, corresponding to our two hands! This example is similar to *although* and *though*, but it merely joins the two statements.)

nevertheless   BDF methods are much less well known than Runge–Kutta methods. *Nevertheless*, there is great demand for BDF codes.

(The second statement is true even though the first statement is true. It would not be correct to replace *however* by *nevertheless* in the previous example, although these two words are sometimes interchangeable.)

despite/in spite of   *Despite* (*in spite of*) the stiff nature of the differential equations that arise in his chemical reaction problems, Professor X prefers to use his favourite Runge–Kutta code.

(Even though his differential equations are stiff, Professor X uses his Runge–Kutta code.)

instead of/rather than For stiff differential equations we use BDF methods *instead of* (*rather than*) Runge–Kutta methods. *Instead of* using (*rather than* use) our usual Runge–Kutta code we turned to a BDF code because we thought the differential equations might be stiff.

(*Rather than* and *instead of* are generally interchangeable but their meaning is different: *rather than* implies a conscious choice, whereas *instead of* merely states an alternative.)

Conditions

Statement a: The indefinite integral does not have a closed form solution.
Statement b: Numerical integration provides an approximation to the definite integral.

if *If* the integral cannot be evaluated in closed form, numerical integration should be used to obtain an approximation.

unless *Unless* the integral can be evaluated in closed form, numerical integration should be used to obtain an approximation.

(Using the converse of the logical condition for *if*.)

whether or not *Whether or not* a closed form exists, numerical integration will always provide an approximation to the integral.

(A numerical approximation can be obtained independent of whether a closed form exists for the integral.)

provided (providing) that *Provided* (*providing*) *that* a closed form solution does not exist, the student may resort to numerical integration.

(More restrictive than *if*. The *that* following *provided* or *providing* can often be omitted, as it can in this sentence.)

Emphasis

Statement: Gaussian elimination with row interchanges does not break down; a zero pivot is a welcome event for it signals that the column is already in triangular form and so no operations need be performed during that step of the reduction.

indeed   Gaussian elimination with row interchanges does not break down; *indeed,* if a zero pivot occurs it signals that the column is already in triangular form and so no operations need be performed during that step of the reduction.
(Introducing a further observation that builds on the first one.)

actually A zero pivot in Gaussian elimination with row interchanges is *actually* a welcome event, for it signals that the column is already in triangular form and so no operations need be performed on that step of the reduction.
(Here, *actually* emphasizes the truth of the statement. It could be replaced by *indeed,* but in the first example *indeed* could not be replaced by *actually;* this is because both words have several slightly different meanings.)

clearly   A zero pivot in Gaussian elimination with row interchanges signals that the column is already in triangular form and that no operations need be performed on that step of the reduction, so the method *clearly* (*obviously, certainly*) cannot break down.

5.6. Spelling

Many English words have alternative spellings. These alternatives fall broadly into two classes. The first class contains those words that are spelled differently in British and American English. The list below illustrates some of the main differences. A special case worthy of mention is the informal abbreviation for mathematics: maths (British English), math (American English).

British spelling	American spelling
behaviour	behavior
catalogue	catalog
centre	center
grey	gray
marvellous	marvelous
skilful	skillful
speciality	specialty
travelling	traveling

The second class of words is those that have alternative spellings in British English.

embed, imbed

learnt, learned. The past tense of the verb *learn.* The advantage of *learnt* is that it avoids confusion with the other meaning of *learned,* which is "having much knowledge acquired by study" (and which is pronounced differently from the first meaning).

spelt, spelled. The past tense of the verb *spell.*

In the last two examples, the -ed ending is the form used in American English. Other examples are

acknowledgement	acknowledgment
encyclopaedia	encyclopedia
focused	focussed
judgement	judgment

There is a host of words, mostly verbs, that can take an -ise or -ize ending. Examples are criticize and minimize. The -ize ending is used in American English, while the -ise ending tends to be preferred in British English (even though *The Concise Oxford Dictionary* gives prominence to the -ize form). A number of words, including the following ones, take only an -ise ending:

advise, arise, comprise, compromise, concise, devise, disguise, exercise, expertise, likewise, otherwise, precise, premise, reprise, revise, supervise, surmise, surprise, treatise.

Verbs ending in -yse (such as *analyse* and *catalyse*) are spelt -yse in British English and -yze in American English.

Several plurals have alternative spellings:

appendices	appendixes
formulae	formulas
indices	indexes
lemmas	lemmata
thesauri	thesauruses
vertices	vertexes

There are also pairs of words that have different meanings in British English but for which one of the pair is often (or always) used with both meanings in American English. Examples: ensure and insure (insure is frequently used for ensure in American English), practice and practise (see page 38). These differences are subtle, and mastering them is not vital to producing clear and effective prose (native speakers also find them confusing). For a good explanation of the reasons for the often haphazard spelling of English, and the reasons for the differences between British and American spelling, I recommend the book by Bryson [30].

Finally, it is important to be aware of words that have very similar spellings or pronunciations, but different meanings. Examples:

accept (agree to receive)	except (but, excluding)
adapt (modify)	adopt (take up, accept)
advice (noun)	advise (verb)
affect (verb: influence)	effect (noun: result) (see also page 37),
complement (the rest)	compliment (flattering remark)
dependant (noun)	dependent (adjective)
device (noun: scheme)	devise (verb: invent)
discreet (careful)	discrete (not continuous)
precede (go before)	proceed (continue)
principal (main)	principle (rule)
stationary (fixed)	stationery (materials for writing)

5.7. Keeping It Simple

The best way to avoid making errors is to keep your writing simple. Use short words and sentences and avoid complicated constructs. Such writing may not be elegant, but it is likely to be unambiguous and readily understood. As your knowledge of English improves you can be more ambitious with your sentence structure and vocabulary. Here is an example giving simple and complicated versions of the same paragraph.

Simple: We note that if the transformation matrix H_p has large elements then A_p is likely to have much larger elements than A_{p-1}. Therefore we can expect large rounding errors in the pth stage, in which case the matrix F_p will have large norm.

Complicated: This conclusion is reinforced if we observe that a transformation with a matrix H_p having large elements is likely to give an A_p with much larger elements than A_{p-1}, and the rounding errors in this stage are therefore likely to be comparatively large, leading to an F_p with a large norm.

5.8. Using a Dictionary

Buy a good monolingual English dictionary and a thesaurus (preferably hardback ones, as you will be using them a lot!). Most bilingual dictionaries do not provide enough detail or wide enough coverage for the advanced writer of English. Also, using a monolingual dictionary will help you to think in English. Instead of, or in addition to, a general-purpose dictionary, you may want to acquire a dictionary written for advanced learners of English, of which there are several. These dictionaries have several notable features. They

- describe a core vocabulary of contemporary English;

- use simple language in their definitions (the *Longman's Dictionary* mentioned below uses a special defining vocabulary of 2000 words);

- give guidance on grammar, style and usage;

- provide examples illustrating typical contexts;

- show pronunciation;

- in some cases, show allowable places to divide a word when it must be split at the end of a line. (This information can also be found in special-purpose dictionaries of spelling and word division.)

Three such dictionaries described as "outstanding" by Quirk and Stein [175] are the *Longman Dictionary of Contemporary English* [136], *Collins Cobuild English Language Dictionary* [44] and the *Oxford Advanced Learner's Dictionary of Current English* [156].

Whereas a dictionary provides meanings for a word, a thesaurus lists alternatives for it that have approximately the same meaning (synonyms). Most thesauri are arranged alphabetically, like a dictionary. In preparation for using your dictionary and thesaurus you should learn the abbreviations they use (these are usually listed at the front) and make sure you understand the grammatical terms noun, adjective, adverb and (transitive or intransitive) verb.

When you are looking for the right word to express an idea, pick whatever words you already know (or that you find in a bilingual dictionary) and look them up in the thesaurus. Then look up in the dictionary the definitions of the synonyms you find and try to decide which is the most appropriate word. This may take some time. It is worth making a note to summarize the search you conducted, as you may later want to retrace your steps.

Watch out for "false friends"—two words in different languages that are very similar but have different meanings. Examples:

French: *location* = the hiring or renting of something. Cf. location.

Italian: *eventuale* = possible. Cf. eventual.

German: *bekommen* = get, receive. Cf. become.

Part of the task of choosing a word is choosing the correct part of speech. Suppose you write, as a first attempt,

The interested Jacobian matrices are those with large, positive dominate eigenvalues.

The two words most likely to be wrong are *interested* and *dominate*, as they can take several different forms. *The Concise Oxford Dictionary* (COD) says dominate is a verb, one meaning of which is "have a commanding influence on". The word we are looking for is an adjective, as it describes a property of eigenvalues. The previous dictionary entry is *dominant*, an adjective meaning most influential or prevailing. This is the correct word. The word *interested* is the correct part of speech: it is an adjective, as it should be since it describes the Jacobian matrices. The COD defines interested as meaning "having a private interest; not impartial or disinterested". It is therefore incorrect to talk about an *interested Jacobian matrix*. The word we require is *interesting*, another adjective meaning "causing curiosity; holding the attention". This example indicates how useful a dictionary can be if you are unsure of vocabulary and grammar. It is a good idea to look up in the dictionary every nontrivial word you write, to check spelling, meaning and part of speech. Do this after you have written a paragraph or section, so as not to interrupt your train of thought.

You will not be able to write perfect English simply by using a dictionary, because even the learner's dictionaries cannot tell you everything you need to know. The best way to learn the subtler aspects of English is to ask someone more fluent in English than yourself to comment on what you have written. If you are not used to writing in English it is almost obligatory to obtain such advice before you submit a paper for publication. Even better is to have a fluent speaker as a co-author. Make sure that you learn from the corrections and suggestions you receive, and keep a note of common mistakes, so as to avoid them.

5.9. Punctuation

There can be differences in punctuation between one language and another. In English, a decimal point separates the integer part of a number from its fractional part ($\pi \approx 3.141$) but in some European languages a comma is used instead ($\pi \approx 3{,}141$). In French, until quite recently, a full stop was used to indicate thousands, so that 2135 would be written 2.135. Now, the official method is to leave a gap: 2 135. (Note that a full stop (British English) is a period (American English).)

Some examples of different sentence punctuation follow, where _____ denotes a sentence or phrase in the given language.

Question: English: _____?
            Greek: _____;
            (romanized) Japanese: _____ ka
            Spanish: ¿_____?

Quotation: English: "_____"
            French, Italian, Spanish: «_____»
            German: „_____" or »_____«
            Swedish: "_____"

Semicolon: English: _____;
            Greek: _____· (but now little used)

5.10. Computer Aids

If you are writing on a computer make regular use of a spelling checker. You may also be able to use a spelling checker or related program to complete or correct a spelling you are unsure of before you put the word into your document. If a style checker is available and you find it helpful, use it too. See §§10.6 and 10.7 for more about spelling checkers and style checkers.

5.11. English Language Qualifications

Two examinations are widely accepted for fulfilling university entrance requirements.

- The Test of English as a Foreign Language (TOEFL) is administered by the Educational Testing Service (required grade 500–580). This test is universally accepted in the US and is becoming more widely accepted in the UK.

- The International English Language Testing System (IELTS) is jointly managed by the British Council, the University of Cambridge Local Examinations Syndicate (UCLES) and the International Development Program of Australian Universities and Colleges (IDP). It is administered by UCLES. This test is widely accepted in the UK and Australia (minimum scores for undergraduate or postgraduate entry range from 4.0 to 7.5) and by an increasing number of institutions in the US.

Ideally, you will have attained these standards before you attempt to write research papers in English. English language courses for non-native speakers are available at many institutions.

5.12. Further Reading

A useful guide on how to use an English dictionary is the booklet by Underhill [211]. It is aimed particularly at users of learner's dictionaries from Oxford University Press and contains many exercises.

Swan's *Practical English Usage* [201] is an invaluable guide for the intermediate or advanced learner of English. It is arranged in dictionary form and covers vocabulary, idiom, grammar, style, pronunciation and spelling. It lists many common examples of incorrect usage, explains differences between British and American English, and distinguishes between formal and informal usage. The *Longman Dictionary of Common Errors* [104] aims to help the reader "avoid or correct those errors most frequently made by foreign learners of English."

Herbert's *The Structure of Technical English* [105] is aimed at students who have completed about six years of learning English and describes "the special structures and linguistic conventions of the English used in technical and scientific writing." This book has been described as pioneering, for being the first textbook in the area now known as English for Specific Purposes (ESP) [200]. Similar books are Swales's *Writing Scientific English* [199] and Ewer and Latorre's *A Course in Basic Scientific English* [59]; the latter contains a useful "dictionary of basic scientific English".

In *Communicating in Science* [27] Booth offers a short chapter titled "Addressed to those for whom English is a foreign language," which includes examples of words that are often used incorrectly.

You may be able to find books on English writing that are designed for readers with your particular linguistic background. For example, I am aware of two excellent books for Swedish speakers. *The Writing Process* [25] is by members of a research group in the English Department at the University of Stockholm and is based on ideas of the Bay Area Writing Project at Berkeley, a widely-used programme for improving writing in American schools and colleges. *Using Numbers in English* by MacQueen [142] explains many aspects of using numbers, including how to say in words expressions such as "1.06", "2115", and "the 1800s"; various uses of "one" and "once"; numbers as adjectives or prepositions; and percentages.

Chapter 6
Writing a Paper

> The title of a book is its special tag, its distinguishing label.
> The choice of words is limited only by the number of words in the English
> language, yet how few of them show up in titles.
>
> — ELSIE MYERS STAINTON, *A Bag for Editors* (1977)

> Go straight to the point, rather than begin with an
> historical reference or resounding banality
> ('There is much research interest at present in the
> biochemistry of memory').
>
> — BERNARD DIXON, *Sciwrite* (1973)

> To write a reference, you must have
> the work you're referring to in front of you.
> Do not rely on your memory. Do not rely on your memory.
> Just in case the idea ever occurred to you, do not rely on your memory.
>
> — MARY-CLAIRE VAN LEUNEN, *A Handbook for Scholars* (1992)

We write scientific papers to communicate our ideas and discoveries. Our papers compete for the readers' attention in journals, conference proceedings and other outlets for scholarly writing. If we can produce well-organized papers that are expressed in clear, concise English, and that convey our enthusiasm and are accessible to people outside our particular speciality, then our papers stand a better than average chance of being read and being referenced. It is generally accepted that the standard of scientific writing is not high[1] [53], [143], [222], so it is not too difficult to produce a paper that stands out from the crowd as being well written.

In this chapter I examine issues specific to writing a paper (or thesis, book, or review), as opposed to the general principles discussed in the previous chapters. After considering the general issues of audience and organization I explore the building blocks of a research paper, from the title to the reference list.

6.1. Audience

Your first task in writing a paper is to determine the audience. You need to identify a typical reader and decide the breadth of your intended readership. Your paper might be written for a mathematics research journal, an undergraduate mathematics magazine, or a book for pre-university students. The formality of the prose and mathematical developments will need to be different in each of these cases. For research journals, at one extreme, you might be writing a very technical paper that builds upon earlier work in a difficult area, and you might be addressing yourself only to experts in that area. In this case it may not be necessary to give much motivation, to put the work in context, or to give a thorough summary and explanation of previous results. At the other extreme, as when you are writing a survey paper, you do need to motivate the topic, relate it to other areas, and explain and unify the work you are surveying. The requirements set forth in the "Guidelines for Authors" for the journal *SIAM Review* are even more specific:

> In their introductory sections, all papers must be accessible to the full breadth of SIAM's membership through the motivation, formulation, and exposition of basic ideas. The importance and intellectual excitement of the subject of the paper must be plainly evident to the reader. Abstraction and specialization must illuminate rather than obscure. The primary threads

---

[1]In 1985, the editor of the British Medical Journal said "The 5000 or so articles we see at the BMJ every year are mostly dreadfully written, with numerous faults in English and overall construction" [135, p. 232].

of the intellectual fabric of the paper can never be hidden by jargon, notation, or technical detail.

The goals described in the last three sentences are worth striving for, whatever your audience.

The audience will determine the particular slant of your paper. A paper about Toeplitz matrices for engineers would normally phrase properties and results in terms of the physical problems in which these matrices arise, whereas for an audience of pure mathematicians it would be acceptable to consider the matrices in isolation from the application.

The language you use will depend on the audience. For example, whereas for linear algebraists I would write about the least squares problem $\min_x \|Ax - b\|$ with its least squares solution x, for statisticians I would translate this to the linear regression problem $\min_b \|Xb - y\|$ and the least squares estimate $\hat{b}$ of the regression coefficients. Failing to use the notation accepted in a given field can cause confusion and can make your work impenetrable to the intended readership.

A good question to ask yourself is why a member of your intended audience should want to read your paper. If the paper is well focused you will find it easy to answer this question. If you cannot find an answer, consider altering the aims of the paper, or doing further work before continuing with the writing.

Whatever your audience, it is worth keeping in mind the words of Ivars Peterson, the editor of *Science News* [166]:

> The format of most journal papers seems to conspire against the broad communication of new mathematical ideas .... The titles, abstracts and introductions of many mathematical papers say: "Outsiders keep out! This is of interest only to those few already in the know."

With a little effort it should be possible to make your work at least partially understandable to non-experts.

6.2. Organization and Structure

At an early point in the writing of your paper you need to think about its high-level organization. It is a good idea to rank your contributions, to identify the most important. This will help you to decide where to put the emphasis and how to present the work, and it will also help you to write the title and abstract.

At the outset you should have some idea of the length of the paper. The larger it is the more important it is that it be well organized. However, it can be harder to write a good short paper than a good long one, for it is difficult to be simultaneously thorough, lucid and concise.

You need to decide in what order to treat topics and how to present results. You should aim to minimize the length, by avoiding repetition; to obtain general results that provide others as special cases (even if the latter are more interesting); and to emphasize similarities and differences between separate analyses.

In addition to considering the first-time reader, you must think about the reader who returns to the paper some time after first studying it. This reader will want to skim through the paper to check particular details. Ideally, then, your paper should not only be easy to read through from beginning to end, but should also function as a reference document, with key definitions, equations and results clearly displayed and easy to find.

6.3. Title

According to Kerkut [116], for every person who reads the whole text of a scientific paper, five hundred read only the title. This statistic emphasizes the importance of the title. The title should give a terse description of the content, to help someone carefully scanning a journal contents page or a reference list decide whether to read the abstract or the paper itself. Ideally, it should also be catchy enough to attract the attention of a browser. Achieving a balance between these two aims is the key to writing an effective title. Kelley [113] mentions that he published an abstract with the title "A decomposition of compact continua and related results on fixed sets under continuous mappings." After Paul Halmos suggested to him that it is not a good idea to put the whole paper into the title, he changed it to "Simple links and fixed sets" for the published paper.

A note on the interpolation problem is too vague a title: what is the breakthrough heralded by *A note*, and which *interpolation problem* is under discussion? Similarly, *Approximation by cubic splines* is too vague (except for a thorough survey of the topic), since the approximation problem being addressed is not clear. Here are some examples of real titles, with my comments.

- *Optimal orthonormalization of the strapdown matrix by using singular value decomposition* [J. MAO, Comput. Math. Appl., 12A (1986), pp. 353–362]. Either *by* or *using* could be omitted.

- *Computing the eigenvalues and eigenvectors of symmetric arrowhead matrices* [DIANNE P. O'LEARY AND G. W. STEWART, J. Comp. Phys., 90 (1990), pp. 497–505]. This is a lively and informative title.

It is good to have action words in the title, such as *computing* or *estimating*.

- *How and how not to check Gaussian quadrature formulae* [WALTER GAUTSCHI, BIT, 23 (1983), pp. 209–216]. "How to" titles immediately arouse the reader's interest.

- *Gaussian elimination is not optimal* [VOLKER STRASSEN, Numer. Math., 13 (1969), pp. 354–356]. If you can summarize your paper in a short sentence, that sentence may make an excellent title.

- *How near is a stable matrix to an unstable matrix?* [CHARLES F. VAN LOAN, in Linear Algebra and its Role in Systems Theory, B. N. Datta, ed., vol. 47 of Contemporary Math., Amer. Math. Soc., 1985, pp. 465–478]. A title that asks a question is direct and enticing.

Punchy titles can be very effective. Here are some examples.

- *ALGOL 68 with fewer tears* [CHARLES H. LINDSEY, Comp. J., 15 (1972), pp. 176–188]. This is an excellent title—it announces that the paper is about the programming language ALGOL 68 and whets the reader's appetite for a demystifying explanation. This paper has the distinction of being a syntactically valid ALGOL 68 program!

- *Nineteen dubious ways to compute the exponential of a matrix* [CLEVE B. MOLER AND CHARLES F. VAN LOAN, SIAM Review, 20 (1978), pp. 801–836]. A classic paper in numerical analysis, with a memorable title.

- *Can you count on your calculator?* [W. KAHAN AND BERESFORD N. PARLETT, Memorandum No. UCB/ERL M77/21, Electronics Research Laboratory, College of Engineering, University of California, Berkeley, Apr. 1977]. Puns rarely make their way into titles, but this one is effective.

- *Performing armchair roundoff analyses of statistical algorithms* [WEBB MILLER, Comm. Statist. B—Simulation Comput., 7 (1978), pp. 243–255]. An otherwise drab title transformed by the word *armchair*. It suggests a gentle approach to the error analysis.

- *Tricks or treats with the Hilbert matrix* [MAN-DUEN CHOI, Amer. Math. Monthly, 90 (1983), pp. 301–312]. This is another attractive and imaginative title.

- *Can one hear the shape of a drum?* [MARC KAC, Amer. Math. Monthly, 73 (4, Part II) Slaught Papers No. 11 (1966), pp. 1–23]. The meaning of this attention-grabbing title is explained on the third page of Kac's expository paper:

  > You can now see that the "drum" of my title is more like a tambourine (which is really a membrane) and that stripped of picturesque language the problem is whether we can determine Ω if we know all the eigenvalues of the eigenvalue problem
  >
  > $$\tfrac{1}{2} \nabla^2 + \lambda U = 0 \quad \text{in} \quad \Omega,$$
  > $$U = 0 \quad \text{on} \quad \Gamma.$$

  This paper won its author the Chauvenet Prize and the Lester R. Ford Award (see Appendix E). Kac's question is answered negatively in *One cannot hear the shape of a drum* [CAROLYN GORDON, DAVID L. WEBB, AND SCOTT WOLPERT, Bull. Amer. Math. Soc., 27 (1992), pp. 134–138] and his "hearing" terminology is used in *You can not hear the mass of a homology class* [DENNIS DETURCK, HERMAN GLUCK, CAROLYN GORDON, AND DAVID WEBB, Comment. Math. Helv., 64 (1989), pp. 589–617].

- *The perfidious polynomial* [JAMES H. WILKINSON, in Studies in Numerical Analysis, Gene H. Golub, ed., vol. 24 of Studies in Mathematics, The Mathematical Association of America, Washington, D.C., 1984, pp. 1–28]. A delightful alliteration for a paper that explains why numerical computations with polynomials can be treacherous. This paper won its author the Chauvenet Prize (see Appendix E).

A few years ago I submitted for publication a manuscript titled "Least squares approximation of a symmetric matrix". A referee objected to the title because it does not fully define the problem, so I changed it to "The symmetric Procrustes problem", which is even less informative if you are not familiar with Procrustes problems, but is perhaps more intriguing.

Unless the title is short it will have to be broken over two or more lines at the head of the paper (or on the front cover of a technical report). Rules of thumb are that a phrase should not be split between lines, a line should not start with a weak word such as a conjunction, and the lines should not differ too much in length. If the title is to be capitalized then all words except articles, short prepositions and conjunctions should be capitalized. Here are some examples, the first of each pair being preferable. The quotes at the beginning of each chapter provide further examples of the choice of line breaks.

On Real Matrices with
Positive Definite Symmetric Component
On Real Matrices with Positive
Definite Symmetric Component

An Iteration Method for the
Solution of the Eigenvalue Problem of
Linear Differential and Integral Operators
An Iteration Method for the Solution of
the Eigenvalue Problem of
Linear Differential and Integral Operators

Numerically Stable Parallel Algorithms
for Interpolation
Numerically Stable Parallel Algorithms for
Interpolation

6.4. Author List

There are no hard and fast rules about the order in which the authors of a multiply authored paper are listed. Sometimes the person who did the greatest part of the work is listed first. Sometimes the academically senior person is listed first. In some disciplines and institutions the senior person (typically the director of a laboratory) is listed last. Perhaps most often, the authors are listed alphabetically, which is the practice I favour.

Being the first-named author is advantageous because the first name is easier to find in a reference list and the paper will be associated solely with that name in citations of the form "Smith et al. (1992)". Also, some citation services ignore all names after the first.

Make sure that you use precisely the same name on each of your publications. I declare myself as Nicholas J. Higham, but not N. J. Higham, N. Higham or Nick Higham. If you vary the name, your publications may not all be grouped together in bibliographic lists and indexes and there may be confusion over whether the different forms represent the same person. Whether to spell out your first name(s) is a matter of personal preference.

6.5. Date

Always date your work. If you give an unpublished paper to others they will want to know when it was written. You may confuse different drafts if they are not dated. The date is usually placed on a line by itself after the author's name, or in a footnote.

The Spotlight Factor

It is the custom in the theoretical computer science community to order authors alphabetically. In a tongue-in-cheek article, Tompa [205] defines the *spotlight factor* of the first author of a paper in which the k authors are listed alphabetically to be the probability that if $k-1$ coauthors are chosen independently at random they will all have surnames later in the alphabet than the first author. According to Tompa, the best (smallest) spotlight factor of 0.0255 in theoretical computer science belongs to Santoro, for her paper NICOLA SANTORO, JEFFREY B. SIDNEY, STUART J. SIDNEY, AND JORGE URRUTIA, *Geometric containment and vector dominance*, Theoret. Comput. Sci., 53 (1987), pp. 345–352. This value is calculated as

$$(1-.\text{santoro})^3 = \left(1 - \left(\frac{19}{27} + \frac{1}{27^2} + \frac{14}{27^3} + \frac{20}{27^4} + \frac{16}{27^5} + \frac{18}{27^6} + \frac{16}{27^7}\right)\right)^3,$$

where $a = 1, \ldots, z = 26$ and blanks or punctuation are represented by zero. By comparison, of those publications in the bibliography of this book, the best spotlight factors are the 0.1829 of O'Leary [154], the 0.2679 of Strunk [198] and the 0.3275 of Knuth [128].

6.6. Abstract

The purpose of the abstract is to summarize the contents of the paper. It should do so in enough detail to enable the reader to decide whether to read the whole paper. The reader should not have to refer to the paper to understand the abstract.

Frequently, authors build an abstract from sentences in the first section of the paper. This is not advisable. The abstract is a mini-paper, and should be designed for its special purpose. This usually means writing the abstract from scratch once the paper is written. The shorter it is, the better, subject to the constraint of it being sufficiently informative. Most abstracts occupy one paragraph. Many mathematics journals state a maximum size for the abstract, usually between 200 and 300 words.

Some specific suggestions are as follows.

- Avoid mathematical equations in the abstract if possible, particularly displayed equations. One reason is that equations may cause difficulties for the review services that publish abstracts (though not those, such as Mathematical Reviews, that use TEX).

- Do not cite references by number in the abstract, since the list of references will not usually accompany the abstract in the review journals. If a paper must be mentioned, spell it out in full:

  An algorithm given by Boyd [*Linear Algebra Appl.*, 9 (1974), pp. 95–101] is extended to arbitrary mixed subordinate matrix norms.

- Try to make the abstract easy to understand for those whose first language is not English. Also, keep in mind that the abstract may be translated into another language, for a foreign review journal, for example.

- Some journals, notably SIAM ones, disallow the word *we* in abstracts, preferring the passive voice (usually necessitating *it*). If you are writing for such a journal it pays to adopt the required voice; while a copy editor will convert as necessary, the conversion may lessen the impact of your sentences.

- Obviously, the abstract should not make claims that are not justified in the paper. Yet this does happen—possibly when the abstract is written before the paper is complete and is not properly revised.

- The abstract should give some indication of the conclusions of the paper. An abstract that ends "A numerical comparison of these methods is presented" and does not mention the findings of the comparison is uninformative.

- Your abstract should lay claim to some new results, unless the paper is a survey. Otherwise, if you submit the paper to a research journal, you are making it easy for a referee to recommend rejection.

- Try not to start the abstract with the common but unnecessary phrases "In this paper" or "This paper". Some journals make this request in their instructions to authors.

The suggestions above are particularly relevant for an abstract that is submitted to a conference and appears in a conference programme. Such an abstract will be read and judged in isolation from the paper, so it is vital for it to create a strong impression in isolation.

An intriguing opening paragraph of an abstract is the one by Knuth (1979) in [122]:

ABSTRACT. Mathematics books and journals do not look as beautiful as they used to. It is not that their mathematical content is unsatisfactory, rather that the old and well-developed traditions of typesetting have become too expensive. Fortunately, it now appears that mathematics itself can be used to solve this problem.

If inspiration fails you, you could always use the following generic abstract from [183].

ABSTRACT. After a crisp, cogent analysis of the problem, the author brilliantly cuts to the heart of the question with incisive simplifications. These soon reduce the original complex edifice to a [s]mouldering pile of dusty rubble.

6.7. Key Words and Subject Classifications

Some journals list key words supplied by the author, usually after the abstract. The number of key words is usually ten or less. Since the key words may be used in computer searches, you should try to anticipate words for which a reader might search and make them specific enough to give a good indication of the paper's content.

Some journals also require subject classifications. The AMS Mathematics Subject Classifications (1991) divide mathematics into 61 sections with numbers between 0 and 94, which are further divided into many subsections. For example, section 65 covers numerical analysis and has 106 subsections; subsection 65F05 covers direct methods for solving linear systems while subsection 65B10 covers summation of series. The classifications are listed in the *Annual Subject Index* of *Mathematical Reviews* and can be downloaded from the American Mathematical Society's e-MATH service (see page 173). Other classification schemes exist, such as the one for computer science from the journal *Computing Reviews*.

The Computing Reviews Classification System (1987) is a four-level tree that has three numbered levels and an unnumbered level of descriptors. The top level consists of eleven nodes, denoted by letters A (General Literature) to K (Computing Milieux). An example of a category is

G.1.3 [Numerical Analysis]: Numerical Linear Algebra—sparse and very large systems.

This specifies the Numerical Linear Algebra node of the Numerical Analysis area under G (Mathematics of Computation), and "sparse and very large systems" is one of several descriptors for G.1.3 listed in the definition of the classification scheme.

6.8. The Introduction

Perhaps the worst way to begin a paper is with a list of notation or definitions, such as *Let G be an abelian group and H be a subgroup of G*, or *Let $\mathcal{F}$ be the complex field $\mathcal{C}$ or the real field $\mathcal{R}$, and let $\mathcal{F}_{m \times n}$ be the linear space of all $m \times n$ matrices over $\mathcal{F}$. If $A \in \mathcal{F}_{m \times n}$ we use A^* to denote the conjugate transpose of A.* It can be argued that the first sentence should be the best in the paper—its job is to entice the uncommitted reader into reading the whole paper. A list of notation will not achieve this aim, but a clear, crisp and imaginative statement may. King [118] gives a slightly different specification for the first sentence: "The first sentence has a dual function: it must carry some essential information, particularly the problem under consideration, and at the same time gently translate the reader into the body of the article."

Ideally, an introduction is fairly short—say a few hundred words. It should define the problem, explain what the work attempts to do and outline the plan of attack. Unless there is good reason not to do so, it is advisable to summarize the results achieved. Knowing the problem and the progress made on it, the reader can decide after reading the introduction whether to read the whole paper. You may want to leave the punch-line to the end, but in doing so you risk the reader losing interest before reaching it.

Here is an example of a compelling opening paragraph that strongly motivates the work. It is from Gautschi's paper "How and How Not to Check Gaussian Quadrature Formulae", mentioned on page 71.

> The preparation of this note was prompted by the appearance, in the chemistry literature, of a 16-digit table of a Gaussian quadrature formula for integration with measure $d\lambda(t) = \exp(-t^3/3)\,dt$ on $(0, \infty)$, a table, which we suspected is accurate to only 1–2 decimal digits. How does one go about convincing a chemist, or anybody else for that matter, that his Gaussian quadrature formula is seriously defective?

It can be appropriate to begin with a few definitions if they are needed to state the problem and begin the analysis. As an example, here is the beginning of the paper "Estimating the Largest Eigenvalue of a Positive Definite Matrix" by O'Leary, Stewart and Vandergraft [154].

> Let A be a positive definite matrix of order n with eigenvalues $\lambda_1 \geq \lambda_2 \geq \cdots \geq \lambda_n > 0$ corresponding to the orthonormal system of eigenvectors $x_1, x_2, \ldots, x_n$. In some applications, one must obtain an estimate of λ_1 without going to the expense of computing the complete eigensystem of A. A simple technique that is applicable to a variety of problems is the power method.

In the next four sentences the authors define the power method, state that the theory of the method is well understood, and note that convergence of the method can be hindered in two ways, which are then analysed. This introduction is effective because it defines the problem and motivates the analysis without a wasted word, and leads quickly into the heart of the paper. The introduction is not labelled as such: this is a three and a half page paper with no section headings.

A possible way to improve an introduction is to delete the first one or more sentences, which are often unimportant general statements. Try it! As an example, consider the following opening two sentences.

> Polynomials are widely used as approximating functions in many areas of mathematics and they can be expressed in various bases. We consider here how to choose the basis to minimize the error of evaluation in floating point arithmetic.

This opening is not a bad one, but the first sentence is general and unexciting. Under the reasonable assumption that the reader knows the importance of polynomials and need not be reminded, it is better to combine and shorten the two sentences and pose a question:

> In which basis should we express a polynomial to minimize the error of evaluation in floating point arithmetic?

Unless the paper is very short it is advisable to outline its organization towards the end of the introduction. One approach is to write, for each section, a sentence describing its contents. The outline can be introduced by a sentence such as "The outline of this paper is as follows" or "This paper is organized as follows." It is best not simply to list the section titles; instead, give a summary that could only be obtained by reading the sections.

6.9. Computational Experiments

Many papers describe computational experiments. These may be done for several reasons: to gain insight into a method, to compare competing methods, to verify theoretical predictions, to tune parameters in algorithms and codes, and to measure the performance of software. To achieve these aims, experiments must be carefully designed and executed. Many decisions have to be made, ranging from what will be the objectives of the experiments to how to measure performance and what test problems to use. One editor of a numerical analysis journal commented that his primary reason for rejecting papers is that the computational experiments are unsound; this underlines the importance of proper design and reporting of experiments. When you report the results of a computational experiment you should give enough detail to enable the results to be interpreted and the experiment to be reproduced. In particular, where relevant you should state the machine precision (working accuracy) and the type of random numbers used (e.g., normal $(0, 1)$ or uniform $(-1, 1)$). If you wish to display the convergence of a sequence it is usually better to tabulate the errors rather than the values themselves. Error measures should be normalized. Thus, for an approximate solution $\widehat{x}$ to $Ax = b$, the relative residual $\|b - A\widehat{x}\|/(\|A\|\|\widehat{x}\| + \|b\|)$ is more meaningful than the scale-dependent quantity $\|b - A\widehat{x}\|$. If you are measuring the speed of a numerical algorithm it is important to show that the right answers are being produced (otherwise the algorithm "the answer is 42" is hard to beat).

You may also wish to state the programming language, the version of the compiler used, and the compiler options and optimizations that were selected, as these can all have a significant influence on run-times. In areas where attaining full precision is not the aim (such as the numerical solution of ordinary differential equations), it is appropriate to consider both speed and accuracy (e.g., by plotting cost versus requested accuracy).

One of the difficulties in designing experiments is finding good test problems—ones which reveal extremes of behaviour, cover a wide range of difficulty, are representative of practical problems, and (ideally) have known solutions. In many areas of computational mathematics good test problems have been identified, and several collections of such problems have been published. For example, collections are available in the areas of nonlinear optimization, linear programming, ordinary differential equations and partial differential equations. Several collections of test matrices are available and there is a book devoted entirely to test matrices. For references to test problem collections see [107].

In your conclusions you should make a clear distinction between objective statements and opinions and speculation. It is very tempting to extrapolate from results, but this is dangerous. As you analyse the results you may begin to formulate conclusions that are not fully supported by the data, perhaps because they were not anticipated when the experiments were designed. If so, further experimentation will be needed. When evaluating numerical algorithms I have found that it pays to print out every statistic that could conceivably be of interest; if I decide not to print out a residual or relative error, for example, I often find a need for it later on.

Further guidelines on how to report the results of computational experiments may be found in the journals *ACM Transactions on Mathematical Software* [47] and *Mathematical Programming* [112], and in an article by Bailey [9]. Perhaps the best advice is to read critically the presentation of experimental results in papers in your area of interest—and learn from them.

6.10. Tables

Many of the principles that apply to writing apply also to the construction of tables and graphs. Three particular points should be considered when designing a table.

(1) It is easier to compare like quantities if they are arranged in columns rather than rows. Research reported by Hartley [100], [102] supports this fact, and Tables 6.1 and 6.2 provide illustration, Table 6.1 being the easier to read. The difference between row and column orientation is more pronounced in complex tables. Of course, the orientation may be determined by space considerations, as a horizontal orientation usually takes less space on the page. If a vertical table is too tall, but is narrow, it can be broken into two pieces side by side:

It is also helpful to put columns or rows that need to be compared next to each other.

(2) Only essential information should be included in a table. Omit data whose presence cannot be justified and state only as many digits as are needed (this number is often surprisingly small). In particular, do not state numerical results to more significant figures than are known for the data. As an example, in Table 6.3, which reports timings for a numerical algorithm on a parallel computer, there is almost certainly no need to quote the timings to six significant figures—two figures would be ample for both the times and the speedups. If you need to present a large amount of data in tabular form, consider displaying it in an appendix, to avoid cluttering the main text. You could give smaller tables in the text that summarize the data. Large sets of data are often better displayed as graphs, however, particularly if it is the trends rather than the numerical values that are of interest. Tufte [206, p. 56] advises that tables are usually better than graphs at reporting small data sets of twenty numbers or less.

(3) To maximize readability the table design should be as simple as possible. In particular, it is best to minimize the number of rules in the table. Two busy examples are shown in Table 6.4 and Table 6.6. The simplified versions, Table 6.5 and Table 6.7, are surely more aesthetically pleasing. Note that there is justification for showing so many digits in Tables 6.4 and 6.5: $\pi(x)$ in this table is the number of primes less than

Multiple	Prefix	Symbol
10^{12}	tera	T
10^9	giga	G
10^6	mega	M
10^3	kilo	K
10^{-1}	deci	d

Table 6.1. SI prefixes $(10^{-1}-10^{12})$. Column orientation.

Multiple	10^{12}	10^9	10^6	10^3	10^{-1}
Prefix	tera	giga	mega	kilo	deci
Symbol	T	G	M	K	d

Table 6.2. SI prefixes $(10^{-1}-10^{12})$. Row orientation.

Number of Processors	Time (secs)	Speedup
1	28.352197	—
4	7.218812	3.9275
8	3.634951	7.7999
16	1.929347	14.6952

Table 6.3. Timings for a parallel algorithm.

or equal to x, and nearly all the digits of $\pi(10^9)$ are needed to show the error in Riemann's formula. Displaying the first one or two digits of the fractional parts of the approximations emphasizes that the approximations are not integers. This table is taken from [123, p. 366],[2] where it is given without any rules and is still perfectly readable.

Give a clear reference to the table at an appropriate place in the text— you cannot rely on the reader to refer to the table automatically. It is helpful if you explain the salient features of the table in words. The reader will appreciate this guidance, especially if the table contains a lot of data. However, you should not summarize the whole table—if you do, the table might as well be omitted.

---

[2]Donald E. Knuth, *The Art of Computer Programming*, vol. 2, ©1981 by Addison-Wesley Publishing Co. Reprinted by permission of Addison-Wesley Publishing Co., Inc., Reading, MA. The form of the original table is not reproduced exactly here.

x	$\pi(x)$	$x/\ln x$	$L(x)$	Riemann's formula
10^3	168	144.8	176.6	168.36
10^6	78498	72382.4	78626.5	78527.40
10^9	50847534	48254942.4	50849233.9	50847455.43

Table 6.4. Approximations to $\pi(x)$.

x	$\pi(x)$	$x/\ln x$	$L(x)$	Riemann's formula
10^3	168	144.8	176.6	168.36
10^6	78498	72382.4	78626.5	78527.40
10^9	50847534	48254942.4	50849233.9	50847455.43

Table 6.5. Approximations to $\pi(x)$.

	Mflops			
n	128	256	512	1024
Method 1 $(n_b = 1)$	95	162	231	283
Method 2 $(n_b = 1)$	114	211	289	330
k variant $(n_b = 1)$	114	157	178	191
Method 1B $(n_b = 64)$	125	246	348	405
Method 2C $(n_b = 64)$	129	269	378	428
k variant $(n_b = 64)$	148	263	344	383

Table 6.6. Results for inverting a lower triangular matrix on a Cray 2.

	n	128	256	512	1024
Unblocked:	Method 1	95	162	231	283
	Method 2	114	211	289	330
	k variant	114	157	178	191
Blocked:	Method 1B	125	246	348	405
$(n_b = 64)$	Method 2C	129	269	378	428
	k variant	148	263	344	383

Table 6.7. Mflop rates for inverting a lower triangular matrix on a Cray 2.

Further Reading

The Chicago Manual of Style [212] devotes a whole chapter to tables and offers much useful advice. Another good reference is *A Manual for Writers* [207, Chap. 6]. Bentley [17, Chap. 10] gives a good example of how to redesign a table. References that discuss the preparation of graphs include [17, Chaps. 10, 11], [100], [102], [140], [141], and [206].

6.11. Citations

The two main styles of citation in mathematics journals are by number (as used in this book) and by name and year, which is the Harvard system. Examples of the Harvard system are *These results agree with an existing study of Smith* (1990) and *These results agree with an existing study (Smith,* 1990). If more than one paper maps to Smith (1990), the papers are distinguished by appending a letter to the year: Smith (1990a), Smith (1990b), and so on. In the number-only system, the number is usually placed in square brackets, though some styles require it to be superscripted.

The main requirement is that a citation does not intrude upon a sentence. For example, *This method was found* [17] *to be unstable* is better written as *This method was found to be unstable* [17]. There are circumstances, however, where a citation has to be placed part-way through a sentence to convey the correct meaning. A good test for whether a citation is well placed is to see whether the sentence reads properly when the citation is deleted. The style of citation inevitably affects how you phrase sentences, so it is worth checking in advance what style is used by the journal in which you wish to publish. Knuth [128] explains that when his paper *"Structured programming with go to statements"* [Computing Surveys, 6 (1974), pp. 261–301] was reprinted in a book, he made numerous changes to make sentences read well with the citation style used in the book.

When you cite by number, it is good style to incorporate the author's name if the citation is more than just a passing one. As well as being more informative, this practice has an enlivening effect because of the human interest it introduces. Examples:

Let $A\Pi = QR$ be a QR factorization with column pivoting [10]. (Passing reference to a textbook for this standard factorization.) The rate of convergence is quadratic, as shown by Wilkinson [27]. (Instead of "as shown in [27]".)

The sentence "This question has been addressed by [5]" is logically incorrect and should be modified to "addressed by Jones [5]" or "addressed in [5]".

When you cite several references together it is best to arrange them so that the citation numbers are in increasing order, e.g., "several variations have been developed [2], [7], [13]." Ordering by year of publication serves no purpose when only citation numbers appear in the text. If you want to emphasize the historical progression it is better to add names and years: "variations have been developed by Smith (1974) [13], Hall (1981) [2], and Jones (1985) [7]."

It is important to be aware that the reference list says a lot about a paper. It helps to define the area in which the paper lies and may be used by a reader to judge whether the author is aware of previous work in the area. Some readers look at the reference list immediately after reading the title, and if the references do not look sufficiently familiar, interesting or comprehensive they may decide not to read further. Therefore it is desirable that your reference list contain at least a few of the key papers in the area in which you are writing. Papers should not, however, be cited just for effect. Each citation should serve a purpose within the paper. Note also that if you cite too often (say, for several consecutive sentences) you may give the impression that you lack confidence in what you are saying.

There are several conventions for handling multiple authors using the Harvard system. One such convention is as follows [33], [51, Chap. 11]. For one or two authors, both names are given (e.g., "see Golub and Van Loan (1989)"). If there are three authors, all three are listed in the first citation and subsequent citations replace the second and third names by "et al." (e.g., "see Knuth, Larrabee and Roberts (1989)", then "see Knuth et al. (1989)"). For four or more authors, all citations use the first author with "et al." These conventions can also be used when naming authors in conjunction with the numbered citation system.

If you make significant use of a result from another reference you should give some indication of the difficulty and depth of the result (and give the author's name). Otherwise, unless readers look up the reference, they will not be able to judge the weight of your contribution.

When you make reference to a specific detail from a book or long paper it helps the reader if your citation includes information that pinpoints the reference, such as a page, section, or theorem number.

For further details on the subtleties of citation consult van Leunen [213].

6.12. Conclusions

If there is a conclusions section (and not every paper needs one) it should not simply repeat earlier sections in the same words. It should offer another viewpoint, discuss limitations of the work, or give suggestions for further research. Often the conclusions are best worked into the introduction or the last section. It is not uncommon to see papers where the conclusions are largely sentences taken from the introduction, such as *We show that X's result can be extended to a larger class* ...; this practice is not recommended.

The conclusions section is a good place to mention further work: to outline open problems and directions for future research and to mention work in progress. Be wary of referring to your "forthcoming paper", for such papers can fail to materialize. A classic example of a justified reference to future work is the following quote from the famous paper by Watson and Crick [218] (*Nature*, April 25, 1953) in which the double helix structure of DNA was proposed:

> It has not escaped our notice that the specific pairing we have postulated immediately suggests a possible copying mechanism for the genetic material.
>
> Full details of the structure, including the conditions assumed in building it, together with a set of co-ordinates for the atoms, will be published elsewhere.

The promise "will be published elsewhere" was fulfilled shortly afterwards, in the May 30, 1953 issue of *Nature*.[3]

6.13. Acknowledgements

Be sure to acknowledge any financial support for your work: grants, fellowships, studentships, sponsorship. A researcher might write "This research was supported by the National Science Foundation under grant DCR-1234567"; grant agencies like authors to be this specific. A Ph.D. student supported by the Science and Engineering Research Council (UK) might write "This work was supported by an SERC Research Studentship." In SIAM journals this type of acknowledgement appears in a footnote on the first page. Other acknowledgements usually appear in a section titled *Acknowledgements* at the end of the paper. (An alternative spelling is *acknowledgments*.)

It is customary to thank anyone who read the manuscript in draft form and offered significant suggestions for improvement (as well as anyone who helped in the research), but not someone who was just doing his or her normal work in helping you (for example, a secretary). The often-used "I would like to thank" can be shortened to "I thank." Note that if you say "I thank X for pointing out an error in the proof of Theorem Y,"

---

[3]Pyke [174] points out that the words "It has not escaped our notice that" can be removed.

you are saying that the proof is incorrect; "in an earlier proof" or "in an earlier attempted proof" is what you meant to say. Finally, don't thank the anonymous referee for *his* suggestions—it may be a *she*.

6.14. Appendix

An appendix contains information that is essential to the paper but does not fit comfortably into the body of the text. The most common use of an appendix is to present detailed analysis that would distract the reader if it were given at the point where the results of the analysis are needed. An appendix can also be used to give computer program listings or detailed numerical results. An appendix should not be used to squeeze inessential information into the paper (though this may be acceptable in a technical report).

6.15. Reference List

Preparing the reference list can be one of the most tedious aspects of writing a paper, although it is made much easier by appropriate software (see Chapter 10). The precise format in which references are presented varies among publishers and sometimes among different journals from the same publisher. Here are four examples.

SIAM journals: J. H. WILKINSON, *Error analysis of floating-point computation*, Numer. Math., 2 (1960), pp. 319–340.

IMA journals: WILKINSON, J. H. 1960 Error analysis of floating-point computation. *Numer. Math.* **2**, 319–340.

Elsevier journals: J. H. Wilkinson, Error analysis of floating-point computation, *Numer. Math.* 2:319–340 (1960).

Springer-Verlag journals: Wilkinson, J. H. (1960): Error analysis of floating-point computation. Numer. Math. **2**, 319–340.

All journals that I am familiar with ask for the use of their own format but will accept other formats and copy edit them as necessary. All publishers have a minimum amount of information that they require for references, as defined in their instructions for authors. It is important to provide all the required information, whatever format you use for the references.

Here are some comments and suggestions on preparing reference lists. For further details I strongly recommend the book by van Leunen [213], but keep in mind that her recommendations may conflict with those of publishers in certain respects.

1. Do not rely on secondary sources to learn the contents of a reference or its bibliographic details—always check the original reference. In studies on the accuracy of citation, the percentage of references containing errors has been found to be as high as 50% [74]. A 1982 paper by J. Vieira and J. Messing in the journal *Gene* had been cited correctly 2,212 times up to 1988, but it had also been cited incorrectly 357 times under "Viera"; these errors led to the paper being placed too low in a list of most-cited papers [77]. In another well-documented case in the medical literature, the Czech title *O Úplavici* ("On Dysentery") of an 1887 paper in a Czech medical journal was taken by one writer to be the author's name, and the mistake propagated until it was finally exposed in 1938 [103], [178]. If a secondary source has to be used (perhaps because the reference is unavailable), it is advisable to append to the reference "cited in [ss]", where [ss] is the secondary source.

2. Always give the full complement of initials of an author. It is not uncommon for only the first initial to be given (perhaps encouraged nowadays by a limitation of BIBTEX—see §10.3).

3. Some authors are inconsistent in the name they use in their papers, sometimes omitting a middle initial, for example. In such cases, my preference is to use the author's full name in the reference list, to avoid ambiguities such as: Is A. Smith the same author as A. B. Smith?

4. Some sources contain typographical errors or nonstandard usage. Titles should be given unaltered. For example, the title "*Van der Monde systems and numerical differentiation*" [J. N. LYNESS AND C. B. MOLER, Numer. Math., 8 (1966), pp. 458–464] appears to be incorrect because the name is usually written Vandermonde, but it should not be altered (I have occasionally had to reinstate "Van der Monde" in my reference list after a copy editor has changed it). A typographical error in an author's name is rare, but not unknown. It seems reasonable to correct such an error, but to provide some indication of the correction that has been made, such as a note at the end of the reference.

5. Copy bibliographic information of a journal article from the journal pages, not the cover of the journal. The cover sometimes contains typographical errors and you cannot deduce the final page number of the article if the journal puts blank pages between articles or begins articles part-way down a page.

6. When you reference a manuscript or technical report that is more than a few months old, check to see if it has appeared in a journal. The author will be happy to inform you of its status. If it has appeared, check whether the title has changed. The referees may ask for a better title, or the copy editor may add a hyphen or combine a hyphenated pair of words.

7. Take care to respect letters and accented characters from other languages. Examples: Å, å, ß, é, ö, ø, Ø.

8. If you maintain a database of references (for example, in BIBTEX format—see §10.3), it is worth recording full details of a reference, even if not all of them are needed for journal reference lists. For a journal article, record the part (issue) number as well as the volume number; this extra information can speed the process of looking up a reference, especially if the journal issues are unbound. For a technical report, the month of publication is useful to know.

9. Be sure you are using the correct journal name and watch out for journals that change their names. For example, the *SIAM Journal on Scientific and Statistical Computing* (1980–1992) became the *SIAM Journal on Scientific Computing* in 1993.

10. In book titles, van Leunen recommends that a colon be added if it is needed to separate a title from a subtitle, and an awkward comma or colon separating a title from a subtitle should be removed. Thus *On Writing Well An Informal Guide to Writing Nonfiction* (as copied from the title page of [225]) needs a colon added after *Well*, and the colon should be removed from *Interpolation Theory:* 5.

11. Van Leunen recommends simplifying the names of major publishers to the bare bones, so that *John Wiley & Sons* becomes *Wiley*, and *Penguin Books* becomes *Penguin*. She also recommends omitting the city for a major publisher; I usually include it because many journals require it. For obscure publishers it is best to give as complete an address as possible.

It is also helpful to give the International Standard Book Number (ISBN) in the more obscure cases. An ISBN consists of ten digits, arranged in four groups whose size can vary. The first group specifies the language group of the publisher ($0, 1$ = English speaking countries, 2 = French speaking, 3 = German speaking, etc.). The second group (2–7 digits) identifies the publisher (e.g., Oxford University Press is 19) and the third group (1–6 digits) identifies the particular title. The last digit is a checksum. If the ISBN is expressed as

$d_1 d_2 \ldots d_{10}$ then

$$d_{10} = \lceil s/11 \rceil * 11 - s, \quad \text{where} \quad s = \sum_{i=1}^{9} (11 - i)d_i$$

($\lceil x \rceil$ denotes the smallest integer greater than or equal to x). A value $d_{10} = 10$ is written as "X". This book has the ISBN 0-89871-314-5: 89871 identifies SIAM as the publisher and 314 is the book's individual number. An International Standard Serial Number (ISSN) identifies a serial publication such as a journal, yearbook or institutional report. An ISSN has eight digits.

12. The date to quote for a book is the latest copyright date (excluding copyright renewals)—ignore dates of reprinting. Always state the edition number if it is not the first.

13. Make sure that every reference is actually cited in the paper. Some copy editors check this, as you may see from their pencilled marks on the manuscript when you receive the proofs.

14. Most mathematics journals require the reference list to be ordered alphabetically by author. Some science journals order by citation, so that the nth paper to be cited is nth in the reference list.

15. A list of standard abbreviations for mathematics journals can be found in *Mathematical Reviews* (see §10.8).

6.16. Specifics and Deprecated Practices

Dangling Theorem

The term *dangling theorem* [113] (or *hanging theorem* [91]) refers to a construction such as the following one, where a theorem dangles from the end of a sentence.

> This result is proved in the following
> **Theorem 3.13.** If f is a twice continuously differentiable function ...

Halmos argues that while the practice can be defended, some readers dislike it, and it is not worth risking annoying them for the sake of avoiding the extra word *theorem*.

The following example does not strictly dangle, but is even more irritating.

5.1. Accuracy of the Computed Solution. It depends on the machine precision and the conditioning of the problem.

Section headings stand alone and should not be taken as part of the text. The obvious solution

5.1. Accuracy of the Computed Solution. The accuracy of the computed solution depends on the machine precision and the conditioning of the problem.

is inelegant in its repetition, but this could be avoided by rewriting the sentence or the title.

Numbering Mathematical Objects

Generally, you should number only those equations that are referenced within the text. This avoids the clutter of extraneous equation numbers and focuses the reader's attention on the important equations. Occasionally it is worth numbering key equations that are not referenced but which other authors might want to quote when citing your paper. Except in very short papers it is best to number equations by section rather than globally (equation (2.3) instead of equation (14)), for this makes referenced equations easier to find. The same applies to the numbering of theorems and other mathematical objects. Whether equation numbers appear on the left or the right of the page depends on the journal.

In referring to numbered equations it is unnecessary to use the word "equation", or an abbreviation, before the equation number. A phrase such as "Now we take norms in (3.9)" is perfectly acceptable, though you might want to say "in equation (3.9)" for emphasis, in certain circumstances. Note that it is strictly incorrect to refer to inequalities, implications or lone expressions as equations.

Two possible numbering sequences are illustrated by

Definition 1, Lemma 1, Theorem 1, Remark 1, Definition 2, Lemma 2, . . .
Definition 1, Lemma 2, Theorem 3, Remark 4, Lemma 5, . . .

Opinions differ as to which is the best scheme. The last has the advantage that it makes it easier to locate a particular numbered item, and the equation numbers themselves can even be included in the sequence for complete uniformity. The disadvantage is that the scheme mixes structures of a different character, which makes it difficult to focus on one particular set of structures (say, all the definitions); and on reading Remark 24 (say), the reader may wonder how many previous remarks there have been. A compromise between the two schemes is to number all lemmas, theorems and corollaries in one sequence, and definitions, remarks and so on in another. Some typesetting systems control the numbering of mathematical objects automatically. LaTeX does so, for example, and the numbering sequence for definitions, lemmas, theorems, etc., can be specified by LaTeX commands.

Plagiarism

Plagiarism is the act of publishing borrowed ideas or words as though they are your own. It is a major academic sin. In writing, if you copy a sentence or more you should either place it in quotes and acknowledge the source via a citation, or give an explicit reference such as "As Smith observed ...." In the case of a theorem statement it is acceptable to copy it word for word if you cite the source, but before copying it you should see whether you can improve the wording or make it fit better into your notation and style.

Regarding when to quote and when to paraphrase, van Leunen [213] advises "Quote what is memorable or questionable, strange or witty. Paraphrase the rest." When you wish to paraphrase, it is best to put the source aside, wait a reasonable period, and then rewrite what you want to say in your own words.

Footnotes

Footnotes are used sparingly nowadays in mathematical writing, and some journals do not allow them (see page 68 for an example of a footnote). It is bad practice to use them to squeeze more into a sentence than it can happily take. Their correct use is to add a note or comment that would deflect from the main message of the sentence. Donald W. Marquardt, the author of the 92nd most-cited paper in the Science Citation Index 1945–1988 [*An algorithm for least-squares estimation of nonlinear parameters, J. Soc. Indust. Appl. Math.*, 11 (1963), pp. 431–441], has stated that a critical part of the algorithm he proposed was described in a footnote and has sometimes been overlooked by people who have programmed the algorithm [76].

Capitalization

References to proper nouns should be capitalized: *See Theorem 1.5, the proof of Lemma 3.4 and the discussion in Section 6.* References to common nouns (generic objects) should not: *Next we prove the major theorems of the paper.*

"This Paper Proves ..."

In the abstract and introduction it is tempting to use wording such as "this paper proves" or "section 3 shows" in place of "we prove" or "we show". This usage grates on the ear of some readers, as it is logically incorrect (though "Theorem 2 gives" cannot be criticized). The grating can be avoided by rewriting, but care is required to avoid a succession of sentences beginning "we".

The Invalid Theorem

Avoid the mistake of calling a theorem into question through sentences such as the following:

> The theorem holds for any continuously differentiable function f.
> Unfortunately, the theorem is invalid because S is not path connected.

A theorem holds and is valid, by definition. A theorem might be *applicable to any continuously differentiable function* or *its invocation* may be *invalid because S is not path connected.*

Chapter 7
Revising a Draft

Simply go through what you have written and
try to curb the length of sentences,
question every passive verb and if possible make it active,
prune redundant words, and look for nouns used instead of verbs.

— BERNARD DIXON, *Sciwrite* (1973)

Every single word that I publish I write at least six times.

— PAUL R. HALMOS, Interview in *Paul Halmos:*
Celebrating 50 Years of Mathematics (1991)

Two pages of the final manuscript . . .
Although they look like a first draft,
they had already been rewritten and retyped—like
almost every other page—four or five times.
With each rewrite I try to make what I have
written tighter, stronger and more precise,
eliminating every element that is not doing useful work.
Then I go over it once more, reading it aloud, and am always
amazed at how much clutter can still be cut.

— WILLIAM ZINSSER, *On Writing Well* (1990)

7.1. How to Revise

All writing benefits from revision. Your first attempt can always be made clearer, more concise, and more forceful. Effective revision is a skill that is acquired through practice. A prerequisite is an understanding of the principles of writing discussed in the previous chapters, together with the ability to spot when these principles have been violated and the skill and courage to rectify the weaknesses.

Put a draft aside for a few days, so that you can examine it with a fresh mind. Then analyse the draft in different ways. Read it aloud and listen to the rhythm. Read it at high speed. Does the text flow? Are the sentences of varying length? Read at the page level, focusing on the shape and the density of ink on the page (assuming the draft has been typeset). Is there a good balance between equations and text? Does the paper look inviting? Do the key ideas stand out? Knuth [128, p. 3] notes that "Many readers will skim over formulas on their first reading of your exposition. Therefore, your sentences should flow smoothly when all but the simplest formulas are replaced by 'blah' or some other grunting noise."

Most good writers work through several drafts—see the quotations at the beginning of the chapter. Some writers adopt the spiral or factorial technique of writing, whereby sections or chapters are written and revised in the order 1, 2, 1, 2, 3, 1, 2, 3, 4, ... Halmos [91] describes this as "the best way to start writing, perhaps the only way." But it is not always necessary to start at the beginning: for example, you can leave the introduction to last and begin by writing a numerical examples section or your conclusions. Every paper is different and versatility is the key.

I make no distinction between editing, revising and rewriting in my own work. I begin with the intention of making whatever changes are necessary to produce the best exposition I can. This may involve minor deletions, insertions and reorderings, or it may require complete rewrites of paragraphs and sections. (As an extreme example, I deleted the original first paragraph of this chapter and started again from scratch—see page 106.) Either way, my printouts usually finish up covered in coloured ink. (Although I made full use of computers in writing this book, I did most of my writing and revising at my desk by writing on a blank sheet or a printout of a draft; only occasionally did I compose at the keyboard. When marking corrections, I use a different colour for each pass through the document.)

Write or print your drafts with wide margins and with a line spacing big enough for you to write between the lines. This will give you a lot of space to mark revisions, which in turn will encourage you to make them.

Here is a short check-list of questions to ask yourself when revising your writing.

▷ Can any words, phrases or sentences be deleted without loss of force or meaning (every such unit must earn its place)?

▷ Can you replace any long, woolly words with short, pithy ones (for example, require → need, utilize → use)?

▷ Are sentences and paragraphs in the best order, or should they be reordered or restructured?

▷ Are there any ambiguities or blatantly wrong words (such as *factor* instead of *feature*)?[a]

▷ Is there any unnecessary repetition?

▷ Can you convert a sentence from the passive to the active voice?

▷ Is every claim fully supported?

▷ Are the mathematical arguments and equations correct?

▷ Have quotations, references and numerical results been copied into the paper correctly?

▷ Is due reference made to the work of other authors (beware of "citation amnesia")?

▷ Are equations, results and the reference list properly numbered? Are all the cross references and bibliographic citations correct?

---

[a] A draft of this book contained the phrase "chop sentences mercifully" instead of "chop sentences mercilessly"!

7.2. Examples of Revision

Many of the points discussed above and in previous chapters are illustrated in the extracts below, most of which are taken from the mathematics and computer science literature. Minor changes have been made to protect the guilty.

▷ GraphX—An experimental graphics development environment research project

> *Comments.*
> This title has the dubious distinction of having five nouns in succession, the first four used adjectivally. It contains three abstract *-al* or *-ent* words. This is not a title to arouse interest.

▷ This type of response surface determination produces information describing not only the variable coordinates for the optimum but also indicates the nature of the response surface in the domain of interest.

> *Comments.*
> The "not only" and "also" clauses do not match. The simplest solution is to delete "indicates". A possible improvement is "The information produced by this type of response surface determination describes not only the variable coordinates for the optimum but also the nature of the response surface in the domain of interest." I would also want to reword the clumsy "response surface determination".

▷ Thus there arises the question of whether a central database can be created in a manner which is consistent with the need for efficiency and breadth of scope.

> *Comments.*
> (1) *Thus there arises the question* can be replaced by the shorter and more direct *The question arises.*
> (2) The which is a wicked one (see page 38). In fact, the "which is" can be deleted.

▷ The Basic Linear Algebra Subprograms (BLAS) have several advantages. They

- Lead to shorter and clearer codes;

- Improve modularity.

- Machine dependent options can be confined to the BLAS, aiding portability, and

- Tuned BLAS have been provided by several manufacturers.

> *Comments.*
> The sense of the bullets is inconsistent, as only the first two match the preceding *they.* The punctuation at the end of the bullets is also inconsistent. The word *options* in the third bullet should be *optimizations.* A tidied up version is: The Basic Linear Algebra Subprograms (BLAS) have several advantages.
> - They lead to shorter and clearer codes.
> - They improve modularity.

- Machine dependent optimizations can be confined to the BLAS, aiding portability.

- Tuned BLAS have been provided by several manufacturers.

▷ *Abstract*—This paper discusses the aims and methods of the FTEsol project and in this context discusses the architecture and design of the control system being produced as the focus of the project.

> *Comments.*
> (1) The phrases *in this context* and *as the focus of the project* serve no useful purpose and can be deleted.
> (2) As mentioned in §6.6, it is not a good idea to begin an abstract with *This paper.* A complete rewrite is needed: The aims and methods of the FTEsol project are discussed, together with the architecture and design of the control system being produced. This version avoids the repeated *discusses* in the original. A more direct alternative, if the use of *we* is allowed, is: We discuss the aims and methods of the FTEsol project . . . .

▷ `Nsolve` finds numerically the 5 complex solutions to this fifth-order equation.

> *Comments.*
> There is a lack of parallelism in *5* and *fifth*, which the reader may find disturbing. I would write "the five complex solutions to this fifth-order equation."

▷ Tables give a systematic and orderly arrangement of items of information. Tabular layout has the particular virtue of juxtaposing items in two dimensions for easy comparison and contrast. Tables eliminate tedious repetition of words, phrases and sentence patterns that can instead be put at the tops of columns and the sides of rows in the table. Although tables do not make much impact by visual display, it is possible, by careful arrangements, to emphasize and highlight particular items or groups of information.

> *Comments.*
> This is the first paragraph of a section on tables from a manual on technical writing. The first sentence reads like a dictionary definition and is surely not telling readers anything they do not know already. The ideas in the third and fourth sentences are not clearly expressed. Revised version:

Tables juxtapose items in two dimensions for easy comparison and contrast. Their row and column labels save the tedious repetition of words that would be necessary if the information were presented in textual form. Although tables lack the visual impact of graphs, information can be grouped and highlighted by careful use of rules and white space.

▷ In order to pinpoint the requirements for an effective microchip development environment sufficiently to definitively obtain answers to the above questions, it is essential to be able to interview a wide variety of microchip developers who are knowledgeable and experienced in such matters.

Comments.

(1) *In order* is superfluous.

(2) *To definitively obtain* is a split infinitive (to-adverb-verb) that is more naturally written as *to obtain definitive answers*. It is probably better to replace the whole phrase by *to answer*.

(3) *Essential to be able to* is perhaps better replaced by *necessary to*.

▷ Due to the advances in computer graphics and robotics, a new interest in geometric investigations has now arisen within computer science focusing mainly on the computational aspects of geometry, forming the research field known as computational geometry.

Comments.

A new interest … has now arisen is a passive phrase that is easily improved. At the same time, we can remove the awkward (and arguably incorrect [198]) *due to*. The phrase *focusing … aspects of geometry* seems unnecessary. My revised version:

Advances in computer graphics and robotics have stimulated a new interest in geometric investigation within computer science, forming the research field known as computational geometry.

▷ In terms of fractals, a straight line has a dimension of one, an irregular line has a dimension of between one and two, and a line that is so convoluted as to completely fill a plane has a dimension approaching the dimension of the plane, namely a dimension of two. Fractal dimensions assign numbers to the degree of convolution of planar curves.

Comments.

This extract is taken from a model paper given in a book about writing a scientific paper. The previous two sentences had also been about fractals. The phrase "in terms of fractals" is unnecessary and the order of these two sentences should be reversed. "Fractal dimensions" do not "assign numbers" but *are* numbers. It is not clear, grammatically, whether the "namely a dimension of two" applies to the convoluted line or the plane. The split infinitive "to completely fill" can be replaced by "to fill". My rewritten version:

> Fractal dimensions are numbers that measure the degree of convolution of planar curves. A straight line has dimension one, an irregular line has a dimension between one and two, and a line that is so convoluted as to fill a plane has a dimension approaching two, which is the dimension of the plane.

▷ Specifically, it is essential to get definitive answers about experiences and judgments concerning specific tool capabilities and the items of information which they utilize and create.

Comments.

(1) Repetition of *specific* is bad style (see §4.16).

(2) The three woolly words *experiences, capabilities* and *utilize* weaken the sentence.

(3) *Items of information* are the same as *information.*

(4) There is a wicked which.

▷ Data flow analysis determines the treatment of every parameter and COMMON variable by every subprogram with sufficient precision that nonportable parameter passing practices can be detected.

Comments.

This sentence is difficult to understand on its first reading. The phrase *by every subprogram* delays the punch-line and the *everys* also delay comprehension. It is not clear whether "with sufficient precision" applies to the subprograms or the analysis. In fact, the latter was intended. A better version is "Data flow analysis determines the treatment of parameters and COMMON variables with sufficient precision that nonportable parameter passing practices can be detected."

▷ Writing $\theta_i = (E|x| + f)_i$ we have

$$\rho(x) \le \| \, |A^{-1}|\mathrm{diag}(\theta_i) \, \|_\infty.$$

Using properties of the ∞-norm we obtain

$$\rho(x) \le \| \, |A^{-1}|(E|x| + f) \, \|_\infty.$$

> *Comments.*
> The presentation hides the fact that the upper bound in the second inequality is equal to that in the first. The following rewrite is more informative:
>
> > Writing $\theta_i = (E|x| + f)_i$, and using properties of the ∞-norm, we obtain
> >
> > $$\begin{aligned} \rho(x) &\le \| \, |A^{-1}|\mathrm{diag}(\theta_i) \, \|_\infty \\ &= \| \, |A^{-1}|(E|x| + f) \, \|_\infty. \end{aligned}$$

▷ [12] reports an eigenproblem from the automobile industry. The eigenvalues of interest are those ones having real part greater than zero.

> *Comments.*
> A citation makes a weak start to a sentence and jolts the eye. In the second sentence *ones* is unnecessary and the sentence can be made more concrete. Revised version:
>
> > Jones reports an eigenproblem from the automobile industry [12]. The eigenvalues of interest are those lying in the right half plane.

▷ Command names have been defined as two letter sequences because it is believed that users prefer to avoid verbosity.

> *Comments.*
> Who is the believer? *It is believed* is better replaced by *we believe*, or *evidence suggests*, or *our experience has shown*, preferably with a reference. A hyphen is needed in *two-letter sequences*, otherwise the meaning is *two sequences of letters*.

▷ When dealing with sets of simple figures, a basic problem is the determination of containment relations between elements of the set.

Comments.

When dealing is a dangling participle: we are not told who is dealing with the sets. A better version results on omitting *when dealing* and changing *the determination of* to *to determine.* Even better is

> A basic problem for sets of simple figures is to determine containment relations between elements of the set.

▷ It is anticipated that the early versions of the system will provide definitive enough information that it will be reasonable to design with some assurance a variety of other systems which should be broader in scope.

Comments.

(1) *Anticipated* should be *expected.* This usage is listed as "disputed" in *The Concise Oxford Dictionary* [158] but is so frequent that it may one day become accepted. Anticipate means to take action against (*The enemy had anticipated our move*).

(2) There are not degrees of definitiveness.

(3) *Reasonable* should perhaps be *possible.*

(4) There is a wicked which. The last phrase could be replaced by *systems of broader scope.*

▷ THEOREM. Let $A \in \mathbb{R}^{m \times n}$ be a given matrix, and let $A = U\Sigma V^T$ be a singular value decomposition of A. Then the problem

$$\max\{Re \text{ trace}(AQ) : Q \in \mathbb{R}^{n \times n} \text{ is orthogonal}\}$$

has the solution $Q = VU^T$, and the value of the maximum is $\sigma_1(A) + \cdots + \sigma_n(A)$, where $\{\sigma_i(A)\}$ is the set of singular values of A.

Comments.

The statement of this theorem is clear and readable but it can be improved. The first sentence can be shortened to "Let $A \in \mathbb{R}^{m \times n}$ have the singular value decomposition $A = U\Sigma V^T$." The second sentence can be rewritten as "Then

$$\max\{Re \text{ trace}(AQ) : Q \in \mathbb{R}^{n \times n} \text{ is orthogonal}\} = \sum_{i=1}^{n} \sigma_i(A),$$

where the $\sigma_i(A)$ are the singular values of A, and the maximum is attained when $Q = VU^T$." This is shorter and, arguably, easier to read. Even better is to define the σ_i at the beginning by appending "where $\Sigma = \text{diag}(\sigma_i)$" to the first sentence.

▷ Let A be $n \times n$. Show that if for any Hermitian matrix H, trace$(HA) = 0$, then $A = 0$.

> **Comments.**
> This question is ambiguous because *any* can mean *whichever* or *at least one* in everyday English. As Halmos [91] recommends, *any* should always be replaced by *each* or *every* in mathematical writing.

▷ Suppose now that the assumption $a = 1$ fails.

> **Comments.**
> An assumption does not fail or succeed: it is either invalid or valid. Better wording might be *Suppose now that the condition $a = 1$ is not satisfied*, or *Suppose now that $a \neq 1$*.

▷ Introduction

Throughout this paper $\|A\|_F$ denotes the Frobenius norm $\left(\sum_{i,j} a_{ij}^2 \right)^{1/2}$ of a real-valued matrix A and A^+ denotes the Moore–Penrose pseudo-inverse of A. We define the set

$$\mathcal{U}_+ = \{ U \in \mathbb{R}^{n \times n} : U^T U = I,\ \det(U) = 1 \}.$$

Given $A, B \in \mathbb{R}^{m \times n}$ we are interested in the *orthogonal Procrustes problem* in its pure rotation form:

$$\min_{U \in \mathcal{U}_+} \|A - BU\|_F.$$

> **Comments.**
> This is a weak start. It can be improved by stating the problem in the first sentence. We can delay the definition of A^+ until it is first used. "Real-valued" can be shortened to "real", but we can dispense with it altogether by using the $\mathbb{R}^{m \times n}$ notation. The "Given" phrase does not read well—interest in the mathematical problem is surely independent of being given matrices A and B. Revised first sentence:
>
> > The pure rotation form of the *orthogonal Procrustes problem* is
> >
> > $$\min_{U \in \mathcal{U}_+} \|A - BU\|_F,$$
> >
> > where
> >
> > $$\mathcal{U}_+ = \{ U \in \mathbb{R}^{n \times n} : U^T U = I,\ \det(U) = 1 \}$$
> >
> > and $\|A\|_F = \left(\sum_{i,j} a_{ij}^2 \right)^{1/2}$ is the Frobenius norm.

The writer should now go on to mention applications in which this problem arises, explain what is known about its solution, and state the purpose of the paper.

▷ This object is achieved by utilizing a set of properties which the signal is known or is hypothesized as possessing.

Comments.
This passive sentence can be shortened and made active:

We achieve this aim by using properties that we know or hypothesize the signal to possess.

This simplified form shows that the sentence says little. The paragraph in which the sentence appears should be rewritten.

▷ THEOREM. Let A be an $n \times p$ complex matrix with rank p. We define the $p \times p$ positive definite Hermitian matrix $S = A^*A$ and the $n \times p$ matrix $\overline{Q} = AS^{-\frac{1}{2}}$. Let $\mathcal{U}$ be the set of all $n \times p$ orthonormal matrices. Then the following is true:

$$\overline{Q} \in \mathcal{U} \qquad \text{and} \qquad \|A - \overline{Q}\| \leq \min_{Q \in \mathcal{U}} \|A - Q\|.$$

Comments.
(1) The rank assumption implies $p \leq n$. It is clearer, though less concise, to say "Let A be an $n \times p$ complex matrix with $n \geq p = \text{rank}(A)$."

(2) There is no need to introduce the symbol S. If the existence of $(A^*A)^{-1/2}$ is thought not to be obvious (note the preferable slashed fraction in the exponent), it can be established in the proof. The symbol $\mathcal{U}$ can also be dispensed with.

(3) The inequality is an equality. The second sentence onwards can be simplified as follows: Let $\overline{Q} = A(A^*A)^{-1/2}$. Then $\overline{Q}$ is orthonormal and

$$\|A - \overline{Q}\| = \min\{ \|A - Q\| : Q^T Q = I_p \}.$$

▷ The optimality of the constant $\pi \log n/4$ in inequality (14.3) is due to Smith [10].

Comments.
This sentence suggests that Smith made the constant optimal. Better is *was first established by Smith.*

▷ Mathematica was found to be a suitable environment in which to perform the computational experiments.

> *Comments.*
> The following rewrite is much shorter, avoids the passive voice, and takes for granted the "suitability": We carried out our computational experiments in Mathematica.

▷ The *main* purpose of any scientific article is to convey in the fewest *number of* words the ideas, procedures and conclusions of an investigator *to the scientific community.* Whether *or not* this *admirable* aim is accomplished depends *to a large extent* on how skillful the author is in *assembling the words of* the *English* language.

> *Comments.*
> These are the opening two sentences of a medical journal editorial titled "Use, Misuse and Abuse of Language in Scientific Writing". The italics are those of Gregory [89], who points out that the italicized words can be omitted without loss of meaning.

▷ (From an abstract:) The bound is derived in the case of k ($0 \leq k \leq p$) explanatory variables measured with error.

> *Comments.*
> The intrusive inequalities and the all-purpose phrase "in the case of" can be removed, and the reader told, or reminded, what p is, by writing "The derivation of the bound allows for any k of the p explanatory variables to be measured with error."

▷ These observations simply imply that nearby orbits separate from the orbit of γ after many iterations of the map G. Hardy and Littlewood (1979) prove a classical theorem that is useful in this context.

> *Comments.*
> There is nothing intrinsically wrong with this extract, but the astute writer may wish to rewrite to remove two minor infelicities. First, the near repetition in the phrase "simply imply" could distract the reader. Second, the eye naturally tends to read "iterations of the map G. Hardy"—symbols can cause trouble at the end of a sentence as well as at the beginning!

Finally, here are some examples from my own writing of how I improved drafts.

(1) Original first two sentences of paper:

▷ Summation of floating point numbers is a ubiquitous and fundamental operation in scientific computing. It is required when evaluating inner products, means, variances, norms, and all kinds of functions in nonlinear problems.

Improved version (shorter, less passive, more direct):

Sums of floating point numbers are ubiquitous in scientific computing. They occur when evaluating inner products, means, variances, norms, and all kinds of nonlinear functions.

An alternative:

Sums of floating point numbers occur everywhere in scientific computing: in evaluating inner products, means, variances, norms, and all kinds of nonlinear functions.

(2) Original:

▷ Here, instead of immediately feeding each correction e_i back into the summation, the corrections are accumulated as $e = \sum_{i=1}^{n} e_i$ (by recursive summation) and then the global correction e is added to the computed sum.

Improved version (omits the unnecessary mathematical notation):

Here, instead of immediately feeding each correction back into the summation, the corrections are accumulated by recursive summation and then the global correction is added to the computed sum.

(3) Original first sentence of abstract:

▷ If a stationary iterative method is used to solve a linear system $Ax = b$ in floating point arithmetic how small can the method make the error and the residual?

Improved version (avoids the misleading *if*, more direct):

How small can a stationary iterative method for solving a linear system $Ax = b$ make the error and the residual in the presence of rounding errors?

(4) Original first sentence of paper:

▷ A *block algorithm* in matrix computations is one that is defined in terms
of operations on submatrices rather than matrix elements.

> A copy editor removed the words *one that is*. This changes the
> meaning, since the sentence now states a property rather than
> gives a definition, but I felt that the shorter sentence was an
> improvement.

(5) Original first paragraph of this chapter.

▷ Perhaps the most important skills you can possess as a writer are the
ability to find faults with your own work and to be able to correct those
faults. It is difficult to look at a draft and make improvements to it, particularly if it is not long since you wrote the draft. This is a skill that is
acquired through practice, but it helps to know what to look for. As you
read a draft you need to keep in mind the principles discussed in the earlier
chapters and never assume that a sentence is beyond improvement. The
examples in this chapter illustrate the type of thinking that is required.

> Note the long, weak opening sentence, which does not contain
> a plural subject for "are".

7.3. A Draft Article for Improvement

Below is a shortened version of an article that I wrote for an undergraduate
mathematics magazine. I have introduced over twenty errors of various
kinds, though most are relatively minor. How many can you spot? If
you are an inexperienced writer, criticizing this "draft" will be a valuable
exercise.

Numerical Linear Algebra in the Sky

In aerospace computations, transformations between different co-ordi-
nate systems are accomplished using the *direction cosine matrix* (DCM),
which is defined as the solution to a time dependent matrix differential
equation. The DCM is 3×3 and exactly orthogonal, but errors in computing
it lead to a loss of orthogonality. A simple remedy, first suggested in a
research paper in 1969 is to replace the computed DCM by the nearest
orthogonal matrix every few steps. These computations are done in real-
time by an aircrafts on-board computer so it is important that the amount

of computation be kept to a minimum. One suitable method for computing a nearest orthogonal matrix is described in this Article. We begin with the case of 1×1 matrices—scalars.

(a) Let x_1 be a nonzero real number and define the sequence:

$$x_{k+1} = \frac{1}{2}(x_k + 1/x_k), \qquad k = 1, 2, \ldots,.$$

If you compute the first few terms on your calculator for different x_1 (e.g. $x_1 = 5$, $x_1 = -3$) you'll find that x_k converges to ± 1; the sign depending on the sign of x_1. Prove that this will always be the case (*Hint:* relate $x_{k+1} \pm 1$ to $x_k \pm 1$ and then divide this two relations). This result can be interpreted as saying that the iteration computes the nearest real number of modulus one to x_1.

(b) This scalar iteration can be generalized to matrices without loosing it's best approximation property. For a given nonsingular $X_1 \in \mathbb{R}^{n \times n}$ define

$$X_{k+1} = \frac{1}{2}(X_k + X_k^{-T}) \qquad (1)$$

(This is one of those very rare situations where it really is necessary to compute a matrix inverse!) Here, X_k^{-T} denotes the transpose of the inverse of X_k. Natural questions to ask are: Is the iteration well defined (i.e., is X_k always nonsingular)? Does it converge? If so, to what matrix?

To investigate the last question suppose that $X_k \to X$. Then X will satisfy $X = \frac{X + X^{-T}}{2}$, or $X = X^{-T}$, or

$$X^T X = I,$$

thus X is orthogonal! Moreover, X is not just any orthogonal matrix. It is the nearest one to X_1 as shown by the following

Theorem 1.

$$\|X - X_1\| = \min\{\|Q - X_1\| : Q \in R^{n \times n}, \ Q^T Q = I_n\}$$

where the norm is denoted by

$$\|X\| = \left(\sum_{i}^{n} \sum_{j=1}^{n} x_{ij}^2 \right)^{1/2}.$$

This is the matrix analogue of the property stated in (a).

Returning to the aerospace application, the attractive feature of iteration (1) is that if we don't wait to long before "re-orthogonalising" our computed iterates then just one or two applications of the iteration (1) will yield the desired accuracy with relatively little work. □

Here are the corrections I would make to the article. (In repeating the exercise myself some time after preparing this section, I could not find all the errors!)

1. First paragraph: hyphenate *time-dependent*; comma after 1969; *aircraft's*; comma after *on-board computer*; *article* in lower case.

2. Second paragraph: no colon after *sequence*. In display, $\frac{1}{x_k}$ instead of $1/x_k$, and replace "...,." by "....". Comma after *e.g.* and instead of semi-colon after ± 1. *These two relations*; *modulus* 1.

3. Third paragraph: *losing*; *its*. Right parenthesis in display (1) is too large and full stop needed at end of display.

4. Fourth paragraph: third X should be in mathematics font, not roman; $(X + X^{-T})/2$. The equation $X^T X = I$ should not be displayed and it should be followed by a semicolon instead of a comma. Comma after X_1; *following theorem*.

5. No need to number the theorem as it's the only one. It should begin with words: *The matrix X_1 satisfies*. $\mathbb{R}^{n \times n}$ instead of R^{nxn}. Comma at end of first display. "I_n" is inconsistent with "I" earlier: make both I. *Denoted* should be *defined*. In first sum of second display, $i = 1$. The parentheses are too large.

6. Last paragraph: *too long*. Wrong opening quotes. For consistency with *generalized* (earlier in article), spell as *re-orthogonalizing*. Logical omission: I haven't shown that the iteration converges, or given or referred to a proof of the theorem.

Chapter 8
Publishing a Paper

In the old days, when table making was a handcraft,
some table makers felt that every entry in a table was a theorem
(and so it is) and must be correct. . . .
One famous table maker used to put in errors deliberately
so that he would be able to spot his work
when others reproduced it without his permission.

— PHILIP J. DAVIS, *Fidelity in Mathematical Discourse:*
Is One and One Really Two? (1972)

The copy editor is a diamond cutter who refines and polishes,
removes the flaws and shapes the stone into a gem.
The editor searches for errors and inaccuracies, and prunes the useless,
the unnecessary qualifiers and the redundancies.
The editor adds movement to the story by substituting active
verbs for passive ones, specifics for generalities.

— FLOYD K. BASKETTE, JACK Z. SISSORS and BRIAN S. BROOKS,
The Art of Editing (1992)

Lotka's law states that the number of people producing n papers
is proportional to $1/n^2$.

— FRANK T. MANHEIM, *The Scientific Referee* (1975)

Memo from a Chinese Economic Journal:
We have read your manuscript with boundless delight.
If we were to publish your paper,
it would be impossible for us to publish any work of lower standard.
And as it is unthinkable that in the next thousand years
we shall see its equal, we are, to our regret,
compelled to return your divine composition,
and to beg you a thousand times to overlook our short sight and timidity.

— From *Rotten Rejections* (1990)

Once your paper has been written, how do you go about publishing it? In this chapter I describe the mechanics of the publication process, from the task of deciding where to submit the manuscript to the final stage of checking the proofs. I do not discuss how to decide whether your work is worth trying to publish, but Halmos offers some suggestions (particularly concerning what not to publish) in [94]. When to publish is an important question on which it is difficult to give general advice. I recommend that you find out the history of some published papers. Authors are usually happy to explain the background to a paper. *Current Contents* (see page 169) regularly carries articles describing the background to a "Citation Classic", which is a paper that has been heavily cited in the literature. A good example is the article by Buzbee [34] describing the story of the paper "On direct methods for solving Poisson's equations" [B. L. BUZBEE, G. H. GOLUB, AND C. W. NIELSON, *SIAM J. Numer. Anal.*, 7 (1970), pp. 627–656]. The article concludes with the following comments.

> So, over a period of about 18 months, with no small amount of mathematical sleuthhounding, we completed this now-*Classic* paper. During that 18 months, we were tempted on several occasions to publish intermediate results. However, we continued to hold out for a full understanding, and, in the end, we were especially pleased that we waited until we had a comprehensive report.

I concentrate here on publishing in a refereed journal. Another important vehicle for publication is conference proceedings. These are more common in computer science than mathematics, and in computer science some conference proceedings are at least as prestigious as journals. It is important to realize that many conference proceedings and a few journals are not refereed, and that when you are considered for hiring or promotion, refereed publications may carry greater weight than unrefereed ones.

8.1. Choosing a Journal

There are more journals than ever to which you can send a scientific paper for publication, so how do you choose among them? The most important question to consider is which journals are appropriate given the content of the paper. This can be determined by looking at recent issues and reading the stated objectives of the journal, which are often printed in each issue. Look, too, at your reference list—any journals that are well represented are candidates for submission of your manuscript. Experts in your area will also be able to advise on a suitable journal.

Several other factors should be considered. One is the prestige and quality of the journal. These rather hard-to-judge attributes depend chiefly on the standard of the papers published, which in turn depends on the standard of the submissions. The higher-quality journals tend to have lower acceptance rates, so publishing in these journals is more difficult. Acceptance rates are usually not published, but may be known to members of editorial boards. The figures sent by SIAM to its editors do not give acceptance rates, but they suggest that the rates for SIAM journals are usually between 30% and 50%, depending on the journal. The prestigious *SIAM Review* has relatively few submissions and an acceptance rate around 50%, but many of these papers are solicited. Gleser [83] states that "the major statistical journals receive many more manuscripts than they can eventually publish and, consequently, have a high rate of rejection," and he remarks that *The Journal of the American Statistical Association* rejects nearly 80% of all papers submitted.

One way to quantify the prestige and quality of a journal is to look at how often papers in that journal are cited in the literature [68], [69]. Such information is provided by the *Journal Citation Reports* published by the Institute for Scientific Information. A study of mathematics journals based on the 1980 report of citation statistics is given by Garfield [71]; his article identifies the 50 most-cited mathematics journals, the most-cited papers from the most-cited journals, and the journals with the highest impact factor (a measure of how often an average article is cited). Based on the 1980 data, the journals with the ten highest impact factors are, from highest to lowest, *Comm. Pure Appl. Math., Ann. Math., Adv. in Math., SIAM Review, Acta Math.—Djursholm, Invent. Math., SIAM J. Numer. Anal., Stud. Appl. Math., Duke Math. J., Math. Program.*

The circulation of a journal should also be considered. If you publish in a journal with a small circulation your paper may not be as widely read as you would like. Relatively new journals from less-established publishers are likely to have small circulations, especially in the light of the budget restrictions imposed on many university libraries. SIAM publishes circulation information in the first issue of the year of each journal; see Table 8.1. The circulation of *SIAM Review* is so high because every SIAM member receives it. If journals are published electronically, as may happen in the coming years, circulation may be less of an issue.

The audience for your paper depends very much on the journal you choose. For example, a paper about numerical solution of large, sparse eigenvalue problems could be published in the *SIAM Journal on Scientific Computing*, where it would be seen by a broad range of workers in scientific computing; in the *SIAM Journal on Matrix Analysis and Applications*, whose readership is more biased towards pure and applied linear algebra; or

Journal	Total Distribution Per Issue, 1991	Issues Per Year
SIAM J. Appl. Math.	2390	6
SIAM J. Comput.	1880	6
SIAM J. Math. Anal.	1318	6
SIAM J. Matrix Anal. Appl.	1155	4
SIAM J. Numer. Anal.	2380	6
SIAM J. Optimization (vol. 1 = 1991)	696	4
SIAM Review	9793	4
SIAM J. Sci. Stat. Comput.	1702	6

Table 8.1. Circulation figures for some SIAM journals.

in the *Journal of Computational Physics*, whose readership is mainly physicists and applied mathematicians, many of whom need to solve practical eigenvalue problems.

Other factors to consider when choosing a journal are the delays, first in refereeing. How long you have to wait to receive referee reports varies among journals. It depends on how much time the journal allows referees, how efficient an editor is at prompting tardy referees, and, of course, it depends on the referees themselves (who usually act as referees for more than one journal).

The other major delay is the delay in publication: the time from when a paper is accepted to when it appears in print. For a particular article, this delay can be calculated by comparing the date of final submission (displayed for each article by most journals) with the cover date of the journal issue. The publication delay depends on the popularity of the journal and the number of pages it publishes each year. A survey of publication delays for mathematics journals appears each year in the *Notices of the American Mathematical Society* journal ("Backlog of Mathematics Research Journals")—it makes interesting reading.

The publication delay also depends, partly, on the author. Ervin Rodin, the editor-in-chief of *Computers and Mathematics with Applications*, explains in an editorial [177] some of the reasons for delays. Four that are not specific to this particular journal are that the figures or graphs are not of high enough quality, the references are not given in full detail, the equations are inconsistently numbered, and the proofs are not returned promptly.

Finally, if your paper has been prepared in TEX (for details of which, see §10.2) you might prefer to send it to a journal that typesets directly from author-supplied TEX source; as well as saving on proofreading this sometimes brings the benefit of extra free reprints (nowadays most journals provide some free reprints for all papers—as few as 15 or as many as 100).

8.2. Submitting a Manuscript

Before submitting your manuscript (strictly, it is a paper only after it has been accepted for publication) you should read carefully the instructions for authors that are printed in each issue of your chosen journal. Most of the requirements are common sense and are similar for each journal. Take particular note of the following points.

1. To whom should the manuscript be submitted? Usually it should be sent to the editor-in-chief, but some journals allow manuscripts to be sent directly to members of the editorial board; judgement is required in deciding whether to take advantage of this option (it may be quicker, since the manuscript skips the stage where an editor-in-chief selects an associate editor). Usually, the editorial addresses are printed in each issue of a journal. (Look at a recent issue, as editors and their addresses can change.)

2. Enclose a covering letter that states to which journal the manuscript is being submitted—this is not obvious if the organization in question has several journals, as does SIAM. State the address for correspondence if there is more than one author. Usually only the designated author receives correspondence, proofs and reprint order forms. If your address will change in the foreseeable future, say so, even if the change is only temporary. (This is particularly important at the typesetting stage, after a paper has been accepted, because proofs must be dealt with quickly.)

3. Most journals request that the manuscript be double-spaced with wide margins, to allow the copy editor room to mark up the text. In LaTeX, when the `article` style is used, a suitable line spacing can be achieved by putting the command `\def\baselinestretch{1.3}` before the `\begin{document}` command.

4. How many copies of the manuscript are required? SIAM requires five. If the destination is abroad, send them by air mail; don't use surface mail, which can take several weeks. Even if a paper is rejected, the manuscript is not usually returned. You should keep a copy, particularly as you may need it at the proofreading stage.

5. Always submit single-sided copies (not double-sided) and fasten them with a staple to avoid pages being lost (the referees can easily remove

the staple if they wish). Provide your full address on the manuscript (some authors forget). Dating the manuscript may help to prick the conscience of a tardy editor or referee.

6. Give key words and the Mathematics Subject Classifications and Computing Reviews classification, if these are required by the journal. It is a good policy to include them as a matter of course. You may also include at this stage a "running head"—a shortened title that appears at the tops of pages in the published paper. The running head should not exceed about 50 characters.

7. If your manuscript cites any of your unpublished work it will help the referees if you enclose a copy of that work, particularly if the manuscript relies heavily on it. Doing so avoids delays that might result from the referees asking to see the unpublished work.

8. If you are using LaTeX double check that the cross references and citations are correct. Adding a new equation or reference at a late stage and not running LaTeX twice (or three times if you are using BibTeX) can result in incorrect numbering. I have seen citation errors of this type persist in a published journal article. Also, if the journal accepts TeX papers, use the style files or macros provided by the publisher; you may still be able to convert the paper to the required format once it has been accepted.

9. You should receive an acknowledgement of receipt of the manuscript within four weeks of submitting it. If you do not, write to ask whether the manuscript was received.

8.3. The Refereeing Process

I will explain how the refereeing process works for SIAM journals (this discussion is partly based on an article by Gear [80]). Procedures for most other journals are similar. If a manuscript is submitted to "The Editor" at the SIAM office, it is logged by SIAM and a letter of acknowledgement is sent to the author, giving a manuscript number that should be quoted in future correspondence. The manuscript is then passed on to the editor-in-chief, who assigns it to a member of the editorial board (this stage can take a few weeks). SIAM (or in some cases, the editor-in-chief) mails the submission, the covering letter, and a Manuscript Transmittal Sheet to the chosen editor. The editor writes to two or more people asking them to referee the paper, suggesting a deadline about six weeks from the time they receive the paper.

The editor may send a sheet of "instructions for referees". Figure 8.1 contains an extract from the SIAM instructions, which indicates again the points to consider before submitting a manuscript.

When all the referee reports have been received the editor decides the fate of the paper, informs the author, and notifies SIAM. (In some journals, including some of those published by SIAM and The Institute of Mathematics and Its Applications, the editor makes a recommendation to the editor-in-chief, who then writes to the author, possibly not naming the editor.) The paper can be accepted, accepted subject to changes, returned to the author for a substantial revision, or rejected.

If a referee report is not received on time the editor reminds the referee that the report is due. After six months of inactivity the manuscript is classed as "flagged" by SIAM, and the editor is urged by SIAM to expedite the current stage of the refereeing process. If you have had no response after six months you are quite entitled to write to the editor-in-chief to enquire whether any progress has been made in refereeing the paper. Papers are sometimes mislaid or lost and your enquiry will at least reveal whether this has happened.

Few papers are accepted in their original state, if only because of minor typographical errors (typos). When preparing a revised manuscript in response to referee reports it is important to address all the points raised by the referees. In the covering letter for the revised version these points might be summarized, with an indication for each one of the action (if any) that was taken for the revision. If you do not act on some of the referees' recommendations you need to explain why. It greatly helps the editor if you explain which parts of the paper have been changed, as it is an irksome task to compare two versions of a paper to see how they differ. When you submit a revised manuscript it is a good idea to mark on the front page "Revised manuscript for journal X" together with the date. This will prevent the revised version from being confused with the original. When a revised paper is received, the editor may ask the referees to look at it or may make a decision without consulting them.

Keep in mind that the editor and referees are usually on your side. They are mostly busy people and would like nothing more than to be able to read your paper, quickly realize that it is correct and deserves publishing, and make that recommendation or decision. Anything you can do to help them is to your benefit. A major advantage of writing in a clear, concise fashion is that your papers may be refereed and edited more quickly!

For those who want to gain further insight into the refereeing process I recommend the papers by Forscher [62], Gleser [83], Manheim [144] and Thompson [204]. Another useful reference is the collection *Publish or Perish* [108], which includes chapters titled "The Refereeing Process", "The Ed-

The most important criterion for acceptance of a publication is originality and correctness. Clear exposition and consistent notation are also required. All papers should open with an introduction to orient the reader and explain the purpose of the paper.

You are asked to prepare an unsigned report, in duplicate. We ask that you keep the report formal and impersonal so that the editor can forward it to the author.

A specific recommendation for acceptance or rejection should be excluded from the report. The following checkpoints are suggested for explicit consideration:

* Is the paper consistent with editorial objectives?

* Is the work correct and original or of wide appeal, if a review paper?

* Is its presentation clear and well organized?

* Is the notation well conceived and consistent?

* How does the paper relate to current literature?

* Are the references complete, relevant, and accurate?

* Does the title accurately characterize the paper?

* Does the abstract properly summarize the paper without being too vague?

* Does the introduction relate the paper to contemporary work and explain the purpose of the paper?

* Are equation numbers and figure numbers consistent?

When the manuscript fails to meet some explicit requirement, what material should the author develop to improve the presentation?

Cover Letter

Please return your report with a cover letter stating your recommendation concerning disposition of the paper. We ask that you justify a recommendation of acceptance as well as one of rejection, and please send the cover letter, report, and manuscript to the editor who requested this review.

Figure 8.1. Extract from SIAM instructions for referees.

Peanuts reprinted by permission of UFS, Inc.

itor's Viewpoint", "The Publisher's Viewpoint", and "Proof-Reading and Correction".

8.4. The Role of the Copy Editor

After your manuscript is accepted for publication it goes to a copy editor. It is important to understand the role of the copy editor. The copy editor of journal papers has three main aims: to do limited rewriting or reorganizing of material in order to make the paper clear and readable; to edit for correctness of grammar, syntax and consistency; and to impose the house style of the journal (a fairly mechanical process). The copy editor tries to make only essential changes and to preserve the author's style.

When you see a copy-marked manuscript for the first time your reaction might be one of horror at the mass of pencilled changes. Most of these will be instructions to the printer to set the paper in the house style: instructions on the fonts to use, spacing in equations, placement of section headings, and a host of other details. There may also be some changes to your wording and grammar. The editor will have had a reason for making each change. The editor will probably have made improvements that you overlooked, such as finding words that are unnecessary, improving the punctuation, and correcting inconsistencies (copy editors keep track of words or phrases that have odd spellings, capitalization or hyphenation and make sure that you use them consistently). If you are unhappy with any of the changes a copy editor has made you can attach an explanatory note to the proof, or, if you are sure an error was made, reverse the change on the proof. Copy editors are always willing to reconsider changes and will pay attention to the author's views in cases of disagreement.

8.5. Checking the Proofs

Some time after your manuscript has been accepted, and after you have received and signed a copyright transferral form, you will receive page or galley proofs (galleys are sheets that have not been broken up into pages—if the journal typesets in TeX the galley stage may be nonexistent). You are asked to check these and return the marked proofs within a short period, often two days. For most journals, the original copy-edited manuscript is enclosed, so that you can see which changes the copy editor marked. If the marked manuscript is not enclosed you are at a disadvantage and you should check against your own copy of the manuscript, particularly for omissions, which can be very difficult to spot when you read only the proofs. The proofs you receive are usually photocopies, so if you see imperfections such as blotches and faint characters these might not be present on the original copy.

A thorough check of the proofs requires one read-through in which you do a line by line comparison with the marked manuscript, and another in which you read the proofs by themselves "for meaning". In addition to checking line by line, do a more global check of the proofs: look at equation numbers to make sure they are in sequence with no omissions, check for consistency of the copy editing and typesetting (typefaces, spacing, etc.) and make sure the running heads are correct. As mentioned above, most journals print the date the manuscript was first received and the date it was accepted. Check these dates, as they may help to establish priority if similar work is published by other researchers.

Some specific errors to check for are shown in Figure 8.2. If your paper contains program listings they should be checked with extra care, as printers find them particularly difficult to typeset. Common errors in typeset program listings include incorrect spacing and indentation and undesired line breaks. It is best to have a listing set from camera-ready copy, if possible.

▷ Unmatched parentheses: $a = (b + c/2$.

▷ Wrong font: $a = (\mathbf{b} + c)/2$.

▷ Missing words or phrases.

▷ Misspelt or wrong words (e.g., complier for compiler, special for spectral).

▷ Repeated words.

▷ Missing punctuation symbols (particularly commas).

▷ Incorrect hyphenation. If your manuscript contains a word hyphenated at the end of a line, the copy editor and printer may not know whether the hyphen is a permanent one or a temporary one induced by the line break.

▷ Widow: short last line of a paragraph appearing at the top of a page. Or "widow word": last word of a paragraph appearing on a line by itself (can be cured in TeX by binding the last two words together with a tie).

▷ O (capital Oh) for 0 (zero), l (lower case ell) for 1, wrong kind of asterisk (* instead of $*$).

▷ Bad line breaks in mathematical equations.

▷ Incorrectly formatted displayed equations (e.g., poor alignment in a multi-line display).

▷ Change in meaning of words or mathematics resulting from copy editor's rewriting.

▷ Missing mathematical symbols (e.g., $\eta \| Lx - b \|$ instead of $\eta = \| Lx - b \|$).

▷ Misplaced mathematical symbols or wrong kind: subscript should be superscript, $\mathbb{R}^{mxn}$ instead of $\mathbb{R}^{m \times n}$, A^t instead of A^T, $\| |A| \|$ instead of $\|A\|$.

▷ Errors in numbers in tables.

▷ Incorrect citation numbers. E.g., if a new reference [4] is added, every citation [n] must be renumbered to [$n + 1$] for $n \geq 4$, but this is easily overlooked.

Figure 8.2. Errors to check for when proofreading.

Copy editors on mathematical journals often have mathematical qualifications, but they may still introduce errors in attempts to clarify, so read very carefully. For example, in one set of proofs that I received the phrase "stable in a sense weaker than" had been changed to "stable, and in a sense, weaker than," which altered the meaning. Since the original may have lacked clarity I changed the phrase to "stable in a weaker sense than." In every set of proofs I have received, I have found mistakes to correct. If you receive apparently perfect proofs, perhaps you are not proofreading carefully enough!

As an indication of how hard it can be to spot typographical errors I offer the following story. Abramowitz and Stegun's monumental *Handbook of Mathematical Functions* [2] was first published in 1964 and various corrections have been incorporated into later printings. As late as 1991, a new error was discovered: the right-hand side of equation (26.3.16) should read $Q(h) - P(k)$ instead of $P(h) - Q(k)$ [180]. This error was spotted when a Fortran subroutine from the NAG library that calculates probabilities behaved incorrectly; the error was traced to the incorrect formula in Abramowitz and Stegun's book.

If you still need convincing of the importance of careful proofreading, consider the book on sky diving that was hurriedly recalled so that an erratum slip could be added. It read "On Page 8, line 7, 'State zip code' should read 'Pull rip cord'".

Corrections should be marked using the standard proofreading symbols; these will be listed on a sheet that accompanies the proofs. A list of proofreading symbols is given in Figure 8.3. In my experience it is possible to deviate slightly from the official conventions as long as the marking is clear. But make sure to mark corrections in the margin, which is the only place a printer looks when examining marked proofs; marks in the text should specify only the location of the correction. Take care to answer any queries raised by the copy editor (usually marked $Au.$ in the margin). (The American Mathematical Society advises its editors to avoid asking "Is it this or that?" because many mathematicians are likely to answer "yes".) Some journals require the printer's errors to be marked in one colour and the author's changes in another. Even if the copy editor doesn't ask, it is advisable to check the reference list to see if any unpublished or "to appear" references can be updated. Don't be afraid of writing notes to the copy editor (I often use yellow stick-on notes), particularly to praise their often under-appreciated work.

You should try to restrict changes you make on the proofs to corrections. If you make other changes you may have to pay for the extra typesetting costs. It is usually possible to add any vital, last-minute remarks (such as a mention of related work that has appeared since you submitted your manuscript) in a paragraph headed "Note Added in Proof" at the end of the paper.

At about the same time as you receive the proofs you will receive a reprint order form and, depending on the publisher, an invitation to pay page charges. Page charges cover the cost of typesetting the article and payment is usually optional for mathematics journals. Even if you request only the free reprints you still need to complete and return the reprint order form.

8.6. Author-Typeset TEX

I now focus on papers that are typeset in TEX by the author (TEX itself is discussed in Chapter 10). SIAM provides macros for use with TEX, LaTEX and AMS-TEX that produce output in the SIAM style. These are available by electronic mail from netlib in the typesetting directory (see §10.9 and the example in §10.10). The macros are also available from SIAM by email or on floppy disk.

In the same way that a computer programmer can write programs that are difficult to understand, an author can produce TEX source that is badly structured and contains esoteric macros, even though it is syntactically correct. Such TEX source is difficult to modify and this can lead to errors being introduced. If you intend to provide TEX source you should make an effort to make it understandable. Watch out for precarious comments, such as those in the following example (the % symbol in TEX signifies that the rest of the line is a comment and should not be printed).

```
The widely used IEEE standard arithmetic
has $\beta=2$. % ANSI/IEEE Standard 754-1985
Double precision has $t=53$, $L=-1021$, $U=1024$,
and $u = 2^{-53} \approx 1.11 \times 10^{-16}$.
% IEEE arithmetic uses round to even.
```

This produces the output

> The widely used IEEE standard arithmetic has $\beta = 2$. Double precision has $t = 53$, $L = -1021$, $U = 1024$, and $u = 2^{-53} \approx 1.11 \times 10^{-16}$.

Suppose that, as these lines are edited on the computer, they are reformatted (either automatically, or upon the user giving a reformatting command) to give

```
The widely used IEEE standard arithmetic has
$\beta=2$. % ANSI/IEEE Standard 754-1985 Double
```

```
precision has $t=53$, $L=-1021$, $U=1024$, and
$u = 2^{-53} \approx 1.11 \times 10^{-16}$. % IEEE
arithmetic uses round to even.
```

As the comment symbols now act on different text, this results in the incorrect output

> The widely used IEEE standard arithmetic has $\beta = 2$. precision has $t = 53$, $L = -1021$, $U = 1024$, and $u = 2^{-53} \approx 1.11 \times 10^{-16}$. arithmetic uses round to even.

Because of this possibility I try to keep comments in a separate paragraph. So I would format the above example as

```
The widely used IEEE standard arithmetic has $\beta=2$.
Double precision has $t=53$, $L=-1021$, $U=1024$,
and $u = 2^{-53} \approx 1.11 \times 10^{-16}$.
```

```
% ANSI/IEEE Standard 754-1985
% IEEE arithmetic uses round to even.
```

Similar difficulties may arise if you edit with a line length of more than 80 characters (the standard screen width on most computers). A different text editor might wrap characters past the 80th position onto new lines, or, worse, truncate the lines; in either case, the meaning of the TeX source is changed.

Errors can be introduced in the transmission of TeX source by email. Characters such as ^, ~, {, } may be interchanged because of incompatibilities between the ASCII character set and other character sets used by certain computers. To warn of translation you could include test lines such as

```
%%   Exclamation    \!    Double quote   \"    Hash (number) \#
%%   Dollar         \$    Percent        \%    Ampersand     \&
%%   Acute accent   \'    Left paren     \(    Right paren   \)
```

at the top of your file.

Some mail systems object to lines longer than 72 characters. Some interpret a line beginning with the word *from* as being part of a mail message header and corrupt the line. Thus

```
from which it follows that
```

might be converted to

```
>from: which it follows that
```

which would be printed in TEX as

¿from: which it follows that

Another possible problem arises when ASCII files are transferred between Unix machines and DOS machines, since Unix terminates lines with a line feed character, whereas DOS uses a carriage-return line-feed pair. Public domain utilities, with names such as dos2unix and unix2dos, are available for converting between one format and the other.

The conclusions from this discussion are twofold.

(1) You should be aware of the potential problems and guard against them. Limit lines to 80 characters (or 72 characters if you will be mailing the file and are unsure which mail systems will be used), keep comment lines separate from the main text, and prepare source that is easy to read and understand.

(2) Read proofs of papers that are typeset from your source with just as much care as those that are re-typeset. Between submission and printing many errors can, potentially, be introduced.

8.7. A SIAM Journal Article—From Acceptance to Publication

What happens once your paper has been accepted for publication in one of SIAM's journals? The following step by step description answers this question. It indicates what you have to do and why the process takes so long. You may be surprised at how many times your paper is checked for errors! This description also explains why you should not expect to be able to make late changes to the paper after you have dealt with the proofs.

TEX or Conventional?

As soon as your manuscript is accepted for publication, SIAM mails you an author-typesetting questionnaire. If you return this questionnaire indicating that a TEX, LATEX, or AMS-TEX file is available, then the journal's production editor will contact you with instructions on submitting your file to SIAM via email. If you have not used the SIAM style macros, which format the paper according to SIAM style specifications, the production editor will tell you how to access them and you will be requested to submit new hard copy of the paper, now formatted with the macros. If a TEX file is not available, then SIAM will have the paper conventionally typeset by a compositor, a process explained below.

Copy Marking

The first process your paper goes through is copy marking (an alternative term for copy editing). The purpose of copy marking is to correct errors in spelling, grammar, punctuation and consistency. When SIAM has not been provided with a TEX source file, it is also necessary to mark the manuscript in such a way that the compositor will typeset it in the SIAM style (which largely follows the guidelines of *The Chicago Manual of Style*). After the paper is copy marked it goes through a quality control stage, which amounts to a second pass of copy marking. If the paper contains any artwork, it is handled at this point. This process usually takes from two to four weeks, depending on the length and complexity of the paper.

Typesetting

Once all the papers for an issue have been copy marked, they are sent in batches to either a compositor or a TEX typesetter. SIAM produces the majority of its TEX papers in-house, but some are sent out to companies that specialize in TEX typesetting.

The compositor keys the manuscript on a computerized typesetting machine and produces galleys. Galleys are printed on resin coated paper commonly referred to as repro. The repro galleys are photocopied onto galley proofs and sent to both you and SIAM for proofreading.

TEX papers are sent, along with disk copies of the TEX files, to the TEX typesetter or in-house TEX specialist. The source files are corrected, using the marked manuscript as a guide, and the corrected files are printed out. These printouts are referred to as galleys, though they do not match the technical definition of a galley. As with conventionally typeset papers, the TEX galleys are mailed to you for proofreading.

Galley Review

Two sets of galleys, with the original copy-marked manuscript and an instruction sheet, are sent directly to you from the compositor or TEX typesetter. You are responsible for proofreading the galleys against the original manuscript and answering all queries. All corrections and answers to questions should be clearly marked on the master set of galleys, and these galleys, along with the original material, should be returned to SIAM. The other set of galleys is for your files. Material for return should be placed in the mail within one week of receipt to ensure that it is included in the assigned issue.

Once you have returned all the material to SIAM, the galleys are checked by a proofreader and the exact length of the article is determined.

Corrections to conventional galleys can be surprisingly expensive. Corrections to TeX galleys can also be expensive as they may cause unexpected changes to the formatting of pages. Try to limit changes to corrections of errors; do not make stylistic changes. The production editor has the authority to cancel expensive or unnecessary changes.

Once all the galleys have been proofread and any queries have been resolved, the final decision on which papers will appear in the issue is made and the order in which they will appear (called the "order of makeup") is determined. At this point you are sent page charge and reprint information, which, in addition to informing you about SIAM's page charge policy, indicates the issue in which your paper will appear. It is possible that the issue assignment that appeared on your galleys will have changed. Often, more articles are set for an issue than necessary, allowing SIAM to have a complete issue even when all galley proofs are not returned on time. It is SIAM's policy that no paper will be published unless the author has returned the galleys.

At this stage, the table of contents is prepared and the journal's frontmatter is updated. The frontmatter comprises all pages preceding the first article page, and usually includes a list of the editorial board and staff of the journal, instructions to authors, and a list of SIAM's sponsors. The galleys and frontmatter are then returned to the compositor or TeX typesetter to be corrected and made into page proofs.

Page Makeup

In convential typesetting, this process consists of pasting up the repro galleys in the proper page depth, placing running heads and folios (page numbers) on each page, and putting artwork into place. These finished pages are photocopied and page proofs are sent to SIAM.

In TeX typesetting, the corrections marked on the galleys are made to the TeX file and the file is printed out again for proofing.

Page Proof Review

The page proof review process is handled at SIAM. Page proofs are checked against the corrected galleys to ensure that all the corrections have been made. TeX papers must be checked at this stage for improper or unsightly reformatting of pages. This can occur automatically when corrections marked on the galleys are incorporated into the TeX files. You are contacted at this point with queries, if necessary, and any late corrections are added. *Only the most critical corrections can be made at the page proof stage.* Corrected pages are returned to the compositor or TeX typesetter.

Once the final corrections are made, the compositor mails the camera-ready repro to SIAM. TeX files are corrected and mailed on floppy disks to a printer, who prints the files out in camera-ready form for SIAM.

Final Repro

This is the last stage before the issue is sent to the printer. Any corrections not previously made are dealt with at SIAM. Once again, only the most critical corrections can be made now because these corrections will be made by hand (by "cutting and pasting"), a difficult process fraught with opportunities for error.

Printing, Binding, and Mailing

Once all corrections have been made, the repro of both the TeX papers and the conventionally typeset papers is sent to the printer. The printer makes negatives from the repro and the negatives are "stripped up", or placed onto printing impositions called flats. A photoprint, or blueline proof, is made from the flats and returned to SIAM. At this stage, the proofs are in journal form, with the cover, frontmatter, text pages, and any advertisements in place. The blueline proofs are checked for dirt, broken type, or the incorrect positioning of pages. This is the last check before printing. Once the blueline is approved the printer makes printing plates from the flats and the issue is printed. After printing, the issue is bound and mailed using SIAM-generated labels.

A Brief History of Scholarly Publishing (extract)

50,000 B.C. *Stone Age publisher demands that all manuscripts be double-spaced, and hacked on one side of stone only.*

1483 *Invention of ibid.*

1507 *First use of circumlocution.*

1859 *"Without whom" is used for the first time in list of acknowledgments.*

1916 *First successful divorce case based on failure of author to thank his wife, in the foreword of his book, for typing the manuscript.*

1928 *Early use of ambiguous rejection letter, beginning, "While we have many good things to say about your manuscript, we feel that we are not now in position . . ."*

1962 *Copy editors' anthem "Revise or Delete" is first sung at national convention. Quarrel over hyphenation in second stanza delays official acceptance.*

—D. D. JACKSON, *The Journal of Irreproducible Results* (1986)

Publication Peculiarities

What is the record for the greatest number of authors of a refereed paper? My nomination is

> P. AARNIO ET AL., *Study of hadronic decays of the Z^0 boson*, Phys. Lett. B, 240 (1990), pp. 271–282. DELPHI Collaboration.

This paper has 547 authors from 29 institutions. The list of authors and their addresses occupies three journal pages. For the shortest titles, I offer

> CHARLES A. MCCARTHY, c_p, Israel J. Math, 5 (1967), pp. 249–271.

and

> NORMAN G. MEYERS AND JAMES SERRIN, $H = W$, Proc. Nat. Acad. Sci. USA, 51 (1964), pp. 1055–1056.

The latter title has the virtue of forming a complete sentence with subject, verb and object. The only title I have seen that contains the word OK is

> THOMAS F. FAIRGRIEVE AND ALLAN D. JEPSON, *O.K. Floquet multipliers*, SIAM J. Numer. Anal., 28 (1991), pp. 1446–1462.

Figure 8.3. Proofreading symbols.

Chapter 9
Writing a Talk

My recommendations amount to this . . .
Make your lecture simple (special and concrete);
be sure to prove something and ask something;
prepare, in detail;
organize the content and adjust to the level of the audience;
keep it short, and, to be sure of doing so,
prepare it so as to make it flexible.

— PAUL R. HALMOS, *How to Talk Mathematics* (1974)

The Three Rules of Public Speaking:
Be forthright.
Be brief.
Be seated.

— SUSAN DRESSEL AND JOE CHEW, *Authenticity Beats Eloquence* (1987)

An awful slide is one which contains approximately a million numbers
(and we've left our opera glasses behind).
An awful lecture slide is one which shows a complete set of
engineering drawings and specifications for a super-tanker.

— KODAK LIMITED, *Let's Stamp Out Awful Lecture Slides* (1979)

9.1. What Is a Talk?

In this chapter I discuss how to write a mathematical talk. By talk I mean a formal presentation that is prepared in advance, such as a departmental seminar or a conference talk, but not one of a series of lectures to students. In most talks the speaker writes on a blackboard or displays pre-written transparencies on an overhead projector. (From here on I will refer to transparencies as slides, since this term is frequently used and is easier to write and say.) I will restrict my attention to slides, which are the medium of choice for most speakers at conferences, but much of what follows is applicable to the blackboard. I will assume that the speaker uses the slides as a guide and speaks freely. Reading a talk word for word from the slides should be avoided; one of the few situations where it may be necessary is if you have to give a talk in a foreign language with which you are unfamiliar (Kenny [115] offers some advice on how to do this).

A talk has several advantages over a written paper [37].

1. Understanding can be conveyed in ways that would be considered too simplified or lacking in rigour for a journal paper.

2. Unfinished work, or negative results that might never be published, can be described.

3. Views based on personal experience are particularly effective in a talk.

4. Ideas, predictions and conjectures that you would hesitate to commit to paper can be explained and useful feedback obtained from the audience.

5. A talk is unique to you—no one else could give it in exactly the same way. A talk carries your personal stamp more strongly than a paper.

Given these advantages, and the way in which written information is communicated in a talk, it is not surprising that writing slides differs from writing a paper in several respects.

1. Usually, less material can be covered in less detail in a talk than in a corresponding paper.

2. Particular care must be taken to explain and reinforce meaning, notation and direction, for a listener is unable to pause, review what has gone before, or scan ahead to see what is coming.

3. Some of the usual rules of writing can be ignored in the interest of rapid comprehension. For example, you can write non-sentences and use abbreviations and contractions freely.

4. Within reason, what you write can be imprecise and incomplete—and even incorrect. These tactics are used to simplify the content of a slide, and to avoid excessive detail. Of course, to make sure that no confusion arises you must elaborate and explain the hidden or falsified features as you talk through the slide.

9.2. Designing the Talk

The first step in writing a talk is to analyse the audience. You need to decide what background material you can assume the listeners already know and what material you will have to review. If you misjudge the listeners' knowledge, they could find your talk incomprehensible at one extreme, or slow and boring at the other. If you are unsure of the audience, prepare extra slides that can be included or omitted depending on your impression as you go through the talk.

The title of your talk should not necessarily be the same one you would use for a paper, because your potential audience may be much different from that for a paper. To encourage non-specialists to attend the talk keep technical terms and jargon to a minimum. I once gave a talk titled "Exploiting fast matrix multiplication within the level 3 BLAS" in a context where non-experts in my area were among the potential audience. I later found out that several people did not attend because they had not heard of BLAS and thought they would not gain anything from the talk, whereas the talk was designed to be understandable to them. A better title would have been the more general "Exploiting fast matrix multiplication in matrix computations."

A controversial title that you would be reluctant to use for a paper may be acceptable for a talk. It will help to attract an audience and you can qualify your bold claims in the lecture. Make sure, though, that the content lives up to the title.

It is advisable to begin with a slide containing your name and affiliation and the title of your talk. This information may not be clearly or correctly enunciated when you are introduced, and it does no harm to show it again. The title slide is an appropriate place to acknowledge co-authors and financial support.

Because of the fixed path that a listener takes through a talk, the structure of a talk is more rigid than that of a paper. Most successful talks follow the time-honoured format "Tell them what you are going to say, say it, then tell them what you said." Therefore, at the start of the talk it is usual to outline what you are going to say: you summarize your objectives and methods, and (perhaps) state your conclusions. This is often done with the aid of an overview slide but it can also be done by speaking over the

---

- Introduction and Motivation
▷ Deriving Partitioned Algorithms
- Block LU factorization and Matrix Inversion
- Exploiting Fast Matrix Multiplication

Figure 9.1. Contents slide: the triangle points to the next topic.

title slide. The aim is to give the listeners a mental road-map of the talk. You also need to sprinkle signposts through the talk, so that the listeners know what is coming next and how far there is to go. This can be done orally (example: "Now, before presenting some numerical examples and my overall conclusions, I'll indicate how the result can be generalized to a wider class of problems"), or you can break the talk into sections, each with its own title slide. Another useful technique is to intersperse the talk with contents slides that are identical apart from a mark that highlights the topic to be discussed next; see Figure 9.1.[1]

Garver [79] recommends lightening a talk by building in multiple entry points, at any of which the listener can pick up the talk again after getting lost. An entry point might be a new topic, problem or method, or an application of an earlier result that does not require an understanding of the result's proof. Multiple exit points are also worth preparing if you are unsure about how the audience will react. They give you the option of omitting chunks of the talk without loss of continuity. A sure sign that you should exercise this option is if you see members of the audience looking at their watches, or, worse, tapping them to see if they have stopped! Whatever the audience's reaction, never overrun the allotted time for the talk.

An unusual practice worth considering is to give a printed handout to the audience. This might help the listeners keep track of complicated definitions and results and save them taking notes, or it might give a list of references mentioned in the talk.

9.3. Writing the Slides

The most common mistake in writing a talk is to put too much on the individual slides. The maxim "less is more" is appropriate, because a busy,

---

[1]To save space, all the boxes that surround the example slides in this chapter are made just tall enough to hold the slide's content.

cluttered slide is hard for the audience to assimilate and may divert their attention from what you are saying. Since a slide is a visual aid, it should contain the core of what you want to say, but you can fill in the details and explanations as you talk through the slide. (If you merely read the slide, it could be argued that you might as well not be there!) There are various recommendations about how many lines of type a slide should contain: a maximum of 7–8 lines is recommended by Kenny [115] and a more liberal 8–10 lines by Freeman et al. [65]. These are laudable aims, but in mathematical talks speakers often use 20 or more lines, somewhat less often to good effect.

A slide may be too long for two reasons: the content is too expansive and needs editing, or too many ideas are expressed. Try to limit each slide to one main idea or result. More than one may confuse the audience and weaken the impact of the points you try to make. A good habit is to put a title line at the top of each slide; if you find it hard to think of an appropriate title, the slide can probably be improved, perhaps by splitting it into two.

When a stream of development stretches over several slides, there is the danger that the audience may wish to refer back to an earlier slide from a later one. To prevent this you can replicate information (an important definition or lemma, say) from one slide to another. A related technique is to build up a slide gradually, by using overlays, or simply by making each slide in a series a superset of the previous one. (The latter effect can be achieved by covering the complete slide with a sheet of paper, and gradually revealing the contents; but be warned that some people find this peek-a-boo style irritating. If you must use this approach cover the slide *before* it goes on the projector, not after.) If you think you will need to refer back to an earlier slide at some particular point, insert a duplicate slide. This avoids the need to search through the pile of used slides. It is worth finding out in advance whether two projectors will be available. If so, you will have less need to replicate material because you can display two slides at a time.

It is imperative to number your slides, so that you can keep them in order at all times. At the end of the talk the slides will inevitably be jumbled and numbers help you to find a particular slide for redisplaying in answer to questions. I put the number, the date of preparation, and a shortened form of the title of the talk, on the header line of each slide.

When you write a slide, aim for economy of words. Chop sentences mercilessly to leave the bare minimum that is readily comprehensible. Here are some illustrative examples.

Original:	It can be shown that $\partial\|Ax\| = \{A^T \text{dual}(Ax)\}$.
Shorter:	Can show $\partial\|Ax\| = \{A^T \text{dual}(Ax)\}$.

Even shorter: $\partial \|Ax\| = \{A^T \mathrm{dual}(Ax)\}$.

Original: The stepsize is stable if $p(z)$ is a Schur polyno-
                 mial, that is, if all its roots are less than one in
                 modulus.

Shorter: Stepsize is stable if $p(z)$ is a Schur polynomial
                 (all roots less than one in modulus).

Even shorter: Stepsize stable if $p(z)$ is Schur ($|\mathrm{roots}| < 1$).

Original: We can perform a similar analysis for the La-
                 grange interpolant.

Shorter: Similar analysis works for the Lagrange inter-
                 polant.

Even shorter: OK for Lagrange.

If you cite your own work you can reduce your name to one letter, but the names of others should not be abbreviated. Thus, from one of my slides:

Original: This question answered in nonsingular case by Higham
           and Knight (1991).

Shorter: Answered for nonsingular A by H & Knight (1991).

Be sure to avoid spelling mistakes on the slides. They are much more noticeable and embarrassing in slides than in a paper.

Graphs and pictures help to break up the monotony of wordy slides. But beware of "slide shuffling", where a long sequence of similar slides is displayed in rapid succession ("that was the plot for $\rho = 0.5$, now here is $\rho = 0.6, \ldots$, and here is $\rho = 0.7, \ldots$"). Anything that might irritate or bore your audience is to be avoided.

At the end of the talk, conclusions and a summary are usually given. This is your last chance to impress your ideas upon the audience. If possible, try to summarize what you have said in a way that does not presuppose understanding of the talk. Even if your conclusions are largely negative, try to finish the talk on a positive note!

Legibility of the Slides

It is of paramount importance that slides be legible when they are projected. Unfortunately, it is a common occurrence that slides are difficult to read for people sitting towards the rear of the room. Various recommendations have been given in the literature on the minimum height of characters, but they necessarily involve assumptions about the magnification of the original image. My advice is to experiment with a projector in an empty room of

---

The Ten Commandments of Giving a Talk

1. Design the talk for the audience.

2. Prepare thoroughly and rehearse the talk.

3. Produce clear, legible slides.

4. Arrive early and check the lecture room.

5. Speak slowly and loudly.

6. Be enthusiastic about what you say.

7. Look at the audience as you speak.

8. Don't fidget with the slides or the pointer.

9. Finish on time.

10. Answer questions courteously and concisely, and admit if you don't know the answer.

---

a similar size to the one that will be used for the talk—you can quickly determine how large the characters need to be. I find that

characters this large (\LARGE in LaTeX at 10pt)

produce a slide readable from the back of most lecture rooms. For the comfort of the audience I prefer to use

characters this large (\huge in LaTeX at 10pt).

For a given character size the font also affects readability [179, Chap. 2]. Some people prefer to use sans serif fonts for slides (fonts without small extensions at the ends of strokes):

sans serif font

An important point to appreciate is that many overhead projector installations exhibit keystoning, whereby the image is wider at the top than at the bottom; this is avoided if the screen is tilted forward. Keystoning causes characters at the bottom of the screen to look much smaller than those at the top.

Take care not to write too close to the edges of a slide, as the areas near the edges may not be clearly visible when the slide is projected. And try to leave a sizeable gap at the bottom of the slide, as the bottom of the screen may be obstructed from the view of people at the back of the audience. Again, I recommend experimenting with a projector to determine acceptable limits.

It is not uncommon to see slides that have been photocopied from a paper without magnification. Invariably the slides are difficult or impossible to read. Avoid this sin: prepare concise, legible slides specially for the talk. The practice of photocopying from a paper is most common with tables. Better even than magnifying the original table is to rewrite it, pruning it to the information essential for that slide. For example, some entries in a table of numbers may be dispensable, and it may be possible to reduce the number of significant digits. More so than in a paper, a table of numerical results may be better displayed as a graph, which is easier to absorb as long as the legends are readable.

Another aspect of writing the talk is making notes and reminders to yourself of what you want to say during each slide. This can be done by writing on the sheets of paper that separate the slides; then you can read the notes as you put the slides onto the projector.

How Many Slides?

The number of slides needed depends on several factors, such as the length of the talk, the amount of mathematical detail to be presented, and the number of graphs and pictures. It also depends on the mode of presentation: a speaker who uses an overhead projector to display outline notes and who spends considerable time filling in the details on the blackboard will need far fewer slides than one who uses only the projector. In my experience, it is best to allow about two minutes per slide for "slide only" talks (the same recommendation is made in [65] and [79]). Allowing for variation in slide length, a 20 minute talk should have a maximum of 12 slides, while a 50 minute talk should not exceed 30 slides.

Handwritten or Typeset?

Which are better: handwritten or typeset slides? It is a matter of personal preference. The best handwritten slides are just as good as the best typeset ones, and the worst of both kinds are very poor. I produce all my slides in LaTeX, for several reasons. I can work from the LaTeX source for the corresponding paper, which saves a lot of typing; typeset slides can easily be adapted and reprinted for use in other talks; I find it easier to revise and correct typeset slides than handwritten ones, and this encourages me to produce better slides; and I have no worries about spilling liquid over my slides or losing them (they are never further away than an ftp call to my workstation). Advantages of handwritten slides are that they can be modified right up until the start of the talk without the need for access to a computer or printer, and it is easy and inexpensive to use colours.

SLiTeX [131, App. A] is a version of LaTeX designed for making slides. It automatically produces large characters in a specially designed sans serif font. It supports the production of coloured slides (each slide is printed as many times as there are colours, with each one containing the output for one colour) and overlays (so that one slide can be superimposed on another). Various other macro packages are available for use with LaTeX, one of the best of which is `seminar.sty` by Timothy Van Zandt (for details, consult `TeX-index`: see page 172). Note that typeset slides are easier to read if they are not right-justified and if hyphenation is turned off.

To break the monotony of black and white typeset slides you can underline headings and key phrases with a coloured pen. You can also leave gaps to be filled during the talk, later wiping the slides clean, ready for the next time you give the talk.

9.4. Example Slides

In this section I give some examples of how to improve slides. In each case two slides are shown: an original draft and a revised version. Keep in mind that the slides would be printed in much larger type, and magnified further by the projector. Therefore, what appear here to be only minor flaws will be much more noticeable in practice.

The slide in Figure 9.2 gives some numerical results. The second version improves on the first in several ways.

1. The definition of the matrix has been changed from a precise but cryptic form to a descriptive one.

2. The quantities ρ^N, ρ^R, and γ_2 would have been defined on an earlier slide, but the audience may need to be reminded of what they represent. Therefore, a description in words has been added.

3. Since only the order of magnitude of the error is relevant, the numbers are quoted to just one significant figure. This reduces the visual complexity of the table. For the same reason, the column headed $\rho^C(\widehat{y})$ has been removed, and the minimum number of rules has been used in the table.

In Figure 9.3, the original slide contains a theorem as it would be stated in a paper. The revised slide omits some of the conditions, which can be mentioned briefly as the slide is explained, and reminds the viewer of the definition of $\kappa_2(A)$. The simplified slide is much easier to assimilate.

The revised slide in Figure 9.5 omits some unnecessary words and symbols from the slide in Figure 9.4. The slide in Figure 9.6 has far too many words. Ruthless pruning produces the much more acceptable version in Figure 9.7; as usual, the information removed from the slide would be conveyed when giving the talk. Note that the parentheses in the expressions for β_{n+1} and π_{n+1} were unnecessary and so have been omitted, and the final error bound is written in a more readable way. Another possible improvement is to remove the definitions of α_0, β_0 and π_0, which may be irrelevant for this slide.

Example

The unit roundoff is $u \approx 1.1921 \times 10^{-7}$.

$$A = \text{randsvd}([10, 16], 1e2)$$

$$\kappa_2(A) = 1.00e2, \text{ cond}_2(A) = 8.63e1$$
$$\text{cond}_2(A, x) = 1.57e2$$

	$\rho^N(\hat{y})$	$\rho^R(\hat{y})$	$\rho^C(\hat{y})$	$\gamma_2(\hat{y})$
Q method	1.83e−8	9.88e−9	1.42e−7	2.01e−6
SNE	5.11e−7	2.79e−7	4.40e−6	4.97e−6
	1.52e−8	6.45e−9	9.64e−8	1.99e−6

Example

Unit roundoff $u \approx 1.2 \times 10^{-7}$.

$$A \quad 10 \times 16, \quad \text{random}$$

$$\kappa_2(A) = 1e2, \quad \text{cond}_2(A) = 9e1$$

	ρ^N	ρ^R	γ_2
Q method	2e−8	1e−8	2e−6
SNE	5e−7	3e−7	5e−6
	2e−8	6e−9	2e−6

ρ^N = normwise backward error
ρ^R = row-wise backward error
γ_2 = forward error

Figure 9.2.

PERTURBATION BOUNDS

Traditional Result

Theorem Let $A \in \mathbb{R}^{m \times n}$ and $0 \neq b \in \mathbb{R}^m$. Suppose rank$(A) = m \leq n$ and that $\Delta A \in \mathbb{R}^{m \times n}$ and $\Delta b \in \mathbb{R}^m$ satisfy

$$\epsilon = \max\{\|\Delta A\|_2 / \|A\|_2, \|\Delta b\|_2 / \|b\|_2\} < \sigma_m(A).$$

If x and $\widehat{x}$ are the minimum norm solutions to $Ax = b$ and $(A + \Delta A)\widehat{x} = b + \Delta b$ respectively, then

$$\frac{\|\widehat{x} - x\|_2}{\|x\|_2} \leq \min\{3, n - m + 2\}\kappa_2(A)\epsilon + O(\epsilon^2).$$

PERTURBATION BOUNDS

$A \in \mathbb{R}^{m \times n}$, $m \leq n$.

$$x = \text{min norm sol'n to } Ax = b,$$
$$\widehat{x} = \text{min norm sol'n to } (A + \Delta A)\widehat{x} = (b + \Delta b).$$

Traditional Bound

$$\frac{\|\widehat{x} - x\|_2}{\|x\|_2} \leq c_{m,n} \, \kappa_2(A)\epsilon + O(\epsilon^2),$$

where

$$\epsilon = \max\left\{\frac{\|\Delta A\|_2}{\|A\|_2}, \frac{\|\Delta b\|_2}{\|b\|_2}\right\},$$

$$\kappa_2(A) = \|A^+\|_2 \|A\|_2.$$

Figure 9.3.

MACHIN'S ARCTAN FORMULA

John Machin (1680–1752) used Gregory's arctan series to
obtain a more rapidly convergent series for computing π. He
argued as follows.

If $\tan\theta = 1/5$, then

$$\tan 2\theta = \frac{2\tan\theta}{1-\tan^2\theta} = \frac{5}{12}$$

and

$$\tan 4\theta = \frac{2\tan 2\theta}{1-\tan^2 2\theta} = \frac{120}{119}.$$

Since $\tan 4\theta = 120/119$ and $\tan(\pi/4) = 1$, this suggests we
evaluate

$$\tan(4\theta - \pi/4) = \frac{\tan 4\theta - 1}{1+\tan 4\theta} = \frac{1}{239}.$$

This gives

$$\arctan\frac{1}{239} = 4\theta - \frac{\pi}{4}$$
$$= 4\arctan\frac{1}{5} - \frac{\pi}{4},$$

or

$$\pi = 16\arctan\frac{1}{5} - 4\arctan\frac{1}{239}.$$

In 1706 Machin used this formula to compute 100 digits of π.

Figure 9.4.

MACHIN'S ARCTAN FORMULA

John Machin (1680–1752) used Gregory's arctan series to obtain a more rapidly convergent series for computing π.

If $\tan\theta = 1/5$, then

$$\tan 2\theta = \frac{2\tan\theta}{1 - \tan^2\theta} = \frac{5}{12}.$$

Similarly, $\tan 4\theta = 120/119$. Then

$$\tan(4\theta - \pi/4) = \frac{\tan 4\theta - 1}{1 + \tan 4\theta} = \frac{1}{239}$$

$$\Rightarrow \quad \arctan\frac{1}{239} = 4\theta - \frac{\pi}{4}$$

$$= 4\arctan\frac{1}{5} - \frac{\pi}{4}.$$

In 1706 Machin computed 100 digits of π.

Figure 9.5.

Fast Methods for Computing π

Up to the 1970s all calculations of π were done using linearly convergent methods based on series, products, or continued fractions. For these, doubling the number of digits in the result roughly doubles the work.

Brent and Salamin independently discovered a quadratically convergent iteration for π in 1976. For such an iteration, the errors satisfy $e_{k+1} = O(e_k^2)$, so each iteration step approximately doubles the number of correct digits.

Borwein and Borwein (1984) discovered another quadratically convergent iteration, and later found even higher order iterations. Their 1984 iteration is

$$\alpha_0 = \sqrt{2}, \quad \beta_0 = 0, \quad \pi_0 = 2 + \sqrt{2},$$

$$\alpha_{n+1} = \frac{1}{2}(\alpha_n^{1/2} + \alpha_n^{-1/2}),$$

$$\beta_{n+1} = \alpha_n^{1/2}\left(\frac{\beta_n + 1}{\beta_n + \alpha_n}\right),$$

$$\pi_{n+1} = \pi_n \beta_{n+1}\left(\frac{1 + \alpha_{n+1}}{1 + \beta_{n+1}}\right),$$

for which $|\pi_n - \pi| \leq \frac{1}{10^{2^n}}$.

Figure 9.6.

Fast Methods for π

Up to 1970s all calculations of π used linearly convergent
series, products, or continued fractions.

Brent and Salamin independently discovered a quadratically
convergent iteration for π in 1976: errors satisfy
$e_{k+1} = O(e_k^2)$.

Borwein and Borwein (1984, 1987) discovered another
quadratically convergent iteration and some higher order
ones. Quadratic iteration: $\alpha_0 = \sqrt{2}$, $\beta_0 = 0$, $\pi_0 = 2 + \sqrt{2}$,

$$\alpha_{n+1} = \frac{1}{2}(\alpha_n^{1/2} + \alpha_n^{-1/2}),$$

$$\beta_{n+1} = \alpha_n^{1/2}\frac{\beta_n + 1}{\beta_n + \alpha_n}, \quad \pi_{n+1} = \pi_n\beta_{n+1}\frac{1 + \alpha_{n+1}}{1 + \beta_{n+1}},$$

for which $|\pi_n - \pi| \leq 10^{-2^n}$.

Figure 9.7.

9.5. Further Reading

In this chapter I have concentrated on *writing* a talk. Excellent advice on both writing and *giving* a talk can be found in various places. In "How to Talk Mathematics", Halmos [93] addresses advice to the young mathematician, but what he says about such topics as simplicity, detail, organization and preparation should be read by all mathematicians (see the quotation at the beginning of this chapter). In *A Handbook of Public Speaking for Scientists and Engineers*, Kenny [115] discusses all aspects of public speaking for the scientist, including preparation of material, choice of visual aids, and presentation. He recommends tongue twisters for the busy scientist to improve articulation! The chapter "Speaking at Scientific Meetings" in Booth's *Communicating in Science* [27] is full of good advice. Turk's *Effective Speaking* [208] is a general guide to speaking; it reports research findings on topics such as audience psychology, memory and non-verbal communication. Calnan and Barabas [37] have written an excellent guide to *Speaking at Medical Meetings*, most of which applies equally well to any form of technical speaking.

Byrne [35] and Garver [79] both offer practical advice on preparing and giving technical presentations. A transcript of a (presumably imaginary) talk that breaks so many rules that it is funny is given in "Next Slide Please" [7]. The American Statistical Association gave workshops on improving technical presentations at its 1980–1982 annual meetings and the material that emerged from the workshops is summarized in "Presenting Statistical Papers" [65]. Common mistakes in using an overhead projector are discussed by Gould in "The Overhead Projector" [85], and in "Visual Aids—How to Make Them Positively Legible" [86] he discusses the choice of letter size for slides, pointing out the effects on legibility of room layout and screen size and positioning. There are many general books on public speaking and much of what they say is relevant to scientific talks. One of my favourites is Aslett's *Is There a Speech Inside You?* [8]. An aspect of public speaking that is often overlooked is proper use of the voice. A useful book in this respect is Berry's *Your Voice and How to Use It Successfully* [22].

Chapter 10
Computer Aids for Writing and Research

$T_{E}X$ can print virtually any mathematical thought that comes into your head,
and print it beautifully.
— HERBERT S. WILF, $T_{E}X$: A Non-review (1986)

I can't compose at a typewriter.
I can't even compose a letter to my relatives on a typewriter,
even though I'm a good typist.
— DONALD E. KNUTH, in Mathematical People:
Profiles and Interviews (1985)

Just as footnotes and a bibliography trace an idea's ancestors,
citation indexing traces an idea's offspring.
— KEVIN KELLY, in SIGNAL: Communication
Tools for the Information Age (1988)

Once you master spelling anonymous,
you can roam around the public storage areas
on computers on the Internet
just as you explore public libraries.
— TRACY LAQUEY and JEANNE C. RYER, The Internet Companion (1993)

10.1. Technical Wordprocessing

Wordprocessing systems have brought a new dimension to writing. Courses are offered on writing with a computer, and books have been written on the subject [149], [223]. Producing mathematical equations is difficult or impossible with wordprocessors not designed for the purpose, but various specially designed systems are available. Because it is ubiquitous in the mathematical and computer science communities I focus here on the TeX typesetting system (the "E X" is pronounced like the *eck* in *deck*). The troff system (with the eqn preprocessor for typesetting mathematics [117]) predates TeX and has been widely used on Unix systems, but it is declining in popularity. My purpose is not to describe or teach TeX, but to explain its advantages and give hints on its use. Gentle introductions for the novice are the short articles by Palais [159] and Wilf [219].

An article by Knuth [122][1] offers many insights into mathematical typesetting and type design, and describes early versions of TeX and META-FONT (METAFONT [125] is Knuth's system for designing typefaces).

A good source of information on TeX is the TeX Users Group (TUG), which runs courses and conferences on TeX and produces a journal called *TUGboat* (beautifully typeset in TeX). It also produces a newsletter called *TeX and TUG News*, which is freely distributed as a LaTeX source file and is available from several archives. For information about joining TUG contact TeX Users Group, P.O. Box 869, Santa Barbara, CA 93102 (email: tug@math.ams.org).

The UK TeX Users Group, based in the UK, also organizes meetings and produces a newsletter (called *Baskerville*). It cooperates with TUG and supports the UKTeX Digest (see §10.9) and a TeX archive at Aston University. For information about joining the UK TeX Users Group contact Graham Toal, 76 Clare Court, Judd Street, London, WC1H 9QW (email: uktug-enquiries@uk.ac.tex).

10.2. TeX

What is TeX?

TeX was developed by Donald Knuth at Stanford University, beginning in the late 1970s [124]. Knuth originally designed TeX to help him typeset books in his *Art of Computer Programming* series. It is a program that accepts a text file interspersed with formatting commands and translates it into a dvi (device independent) file that can readily be translated into the languages understood by output devices such as screens and laser printers.

---

[1]The beginning of the abstract is quoted on page 75.

The typical usage is to prepare an ASCII file using your favourite editor, run it through TEX and preview the output on the screen. You correct errors by editing the source file and then re-running TEX. When you wish to obtain hard copy, you invoke a driver to render the TEX output on the printer.

LATEX [131] by Leslie Lamport and AMS-TEX [185] by Michael Spivak are macro packages that "sit on top" of TEX. I recommend that you use one of these packages instead of TEX with its basic `plain` macros. AMS-TEX is particularly appropriate for those mathematicians who need special fonts or commutative diagrams; it is used by all publications of the American Mathematical Society (AMS). LATEX is widely used in the mathematical sciences and is the most popular choice for producing books.

AMS-LATEX from the AMS [6] brings many features of AMS-TEX to LATEX, but retains the syntax of LATEX. (It can be downloaded from the AMS's e-MATH service; see §10.9.) LAMS-TEX [186], written by Michael Spivak, is a set of macros built on AMS-TEX that adds many LATEX-like features.

There are several advantages to preparing your papers in some form of TEX.

1. TEX is available on many computer systems, from PCs to workstations to mainframes, and output can be obtained on devices ranging from screens to dot matrix printers to professional typesetting machines.

2. Many journals now accept papers in TEX format; this saves the usual typesetting stage and its inevitable introduction of errors.

3. In LATEX and AMS-TEX the format of the document is governed by style options, which are specified at the beginning of the source. Only the style options need to be changed to alter the format of the document. Available formats range from book to technical report to SIAM or AMS journal paper.

4. TEX source is plain ASCII text so it can be sent by electronic mail or ftp'ed (see §10.10). This is convenient when collaborating with a distant co-author or submitting a final paper in TEX format to a journal.

5. TEX's `dvi` files, or their PostScript translations, are relatively portable means for distributing papers electronically. For example, you can make your papers available by anonymous ftp across the Internet network (see §10.10). (Making the `tex` file available is less satisfactory because it is often inconvenient to collect all the source into one file, particularly if LATEX is used with nonstandard style files.)

6. Cross-references to equations, theorems, tables, etc. and citations of bibliographic references are fully supported using symbolic labels in LaTeX. Renumbering of the cross-references is handled automatically when a new equation or bibliographic entry is added.

7. It is easy to prepare transparencies with TeX if the paper on which the talk is based is already in TeX form, as the conversion largely consists of deleting chunks of text and equations. See §9.3.

Hints on Using TeX

In this section I offer some suggestions on the use of TeX, beginning with two generally useful ideas.

When typing your paper make liberal use of spaces and indentation to improve the readability of the source. Start new sentences and main phrases on new lines. This approach simplifies editing: the source is easier to navigate, and cut and paste operations are easier because they tend to act on whole sentences or phrases. It may also be desirable to limit lines to 72 characters, for reasons explained in §8.6. Here is a short extract from a paper of mine (in LaTeX) that illustrates these points:

```
We assume that $x_{k+1}$ is computed by
forming $Nx_k+b$ and then solving a linear system with $M$.
The computed vectors $\xhat_k$ therefore satisfy
\begin{equation}
        (M + \dM_{k+1})\xhat_{k+1} = N\xhat_k + b + f_k,
  \label{be1}
\end{equation}
```

It is a good policy to use macros to typeset important notation, whether it be words or symbols. As well as reducing the amount of typing and ensuring consistency, this helps to emphasize logical structure in the source, which will be appreciated if you have to change the notation, for only the macro definitions will need to be altered. The latter point is discussed by Lamport [132], who argues against WYSIWYG (what you see is what you get) systems and in favour of markup systems such as LaTeX. As an example, among the macros used in typesetting this book are

```
\def\mw{mathematical writing}
\def\norm#1{\|#1\|}
\def\Htk{\widetilde{H_k}}
```

The first macro is used to save typing. (I type \mw instead of "mathematical writing".) I find the second macro makes the source code easier to read and modify. (I type `\norm{x}` to produce $\|x\|$.) It makes it easy to change from a general norm to a specific one if the need arises (e.g., to convert to the 2-norm the definition is changed to `\def\norm#1{\|#1\|_2}`). The third macro is used in the source for §3.5; again, it saves typing and simplifies any future change of notation.

The next group of suggestions relates to typesetting general text.

Ties and Spaces

Avoid unfortunate line-breaks by binding units together with a tie (`~`). For example, as a matter of course type `Theorem~3` not `Theorem 3`, `Mr.~Sandman` not `Mr. Sandman`, and `1,~2 and~3` not `1, 2 and 3`. Other bad breaks, such as between the last two words in a paragraph, can be corrected as and when they appear.

Leave a space before the `\cite` command.

> Bad: `Knuth\cite{knut86}`, giving Knuth[124].
>
> Good: `Knuth \cite{knut86}`, giving Knuth [124].
>
> Good: `Knuth~\cite{knut86}`, giving Knuth [124].

The "Knuth[124]" form is a common mistake. The last form avoids a line break just before the citation.

Remember to put a control space after a full stop if the full stop does not mark the end of the sentence. Thus p. 12 is typed as `p.\ 12`, and cf. Smith (1988) as `cf.\ Smith (1988)`. In the references in a `thebibliography` environment there is no need to use control spaces because this environment redefines the full stop so that it does not give end of sentence spacing.

Dashes

There are three types of dashes in typesetting. A hyphen is produced in TₑX by typing `-`, as in state-of-the-art computer. Number ranges use the en-dash, typed in TₑX as `--` rather than a single dash. This gives 1975–1978 and pp. 12–35 instead of 1975-1978 and pp. 12-35. The en-dash is also used when the first part of a compound word does not modify the meaning of the second part; it can be thought of as standing for *and* or *to*. Examples: Moore–Penrose inverse, left–right evaluation. A dash partitioning a sentence—like this—should be typed as `---` with no space on either side; this is the em-dash. A mathematical minus sign is produced by a single `-` within mathematics mode. Compare -3, typed as `-3`, with -3, typed as `$-3$`.

Quotes, Dates, Lists and Paragraphs

Be careful to distinguish between the opening quote (' or \lq) and the closing quote (' or \rq). Errors such as "unmatched quotes" are irritating to the trained eye. Do not use the double quote ", which always produces the right quotes ".

In LaTeX set the date explicitly using the \date command (assuming you use \maketitle)—otherwise the date that is printed will be the date on which the file was LaTeXed, which is less informative.

Be careful not to overuse lists. LaTeX makes it very easy to produce bulleted or numbered lists, with the itemize and enumerate environments. An inappropriate use of lists is to present a series of ideas without linking them smoothly. A list may be the remnant of an outline that has not been fully developed. If a list is longer than a full page, or has more than one paragraph in an entry, it is worth considering converting it to running text.

It is easy to start new paragraphs unintentionally in TeX by leaving a blank line after a list, quotation or mathematical display. Watch out for this error, as it is easy to overlook when you are revising and proofreading and it can be disconcerting for the reader.

The final group of suggestions relates to typesetting mathematics. Appendix B summarizes the TeX and LaTeX symbols and spacing commands.

Spacing in Formulas

Correct spacing in displayed formulas can usually be accomplished with the \quad and \qquad commands. A \quad is a printer's quad of horizontal space, which is approximately the width of a capital M, and a \qquad is twice the width of a \quad. A \qquad is appropriate when a formula is followed by a qualifying expression. Less often, a thick space, typed as \; (5/18 of a quad), is useful. Here are some examples of spacing.

$$\|A^T\|_p = \|A\|_q, \qquad \frac{1}{p} + \frac{1}{q} = 1$$
$$(A + E)\hat{x} = b, \qquad \|E\| \le \epsilon\|A\|$$
$$fl(x \text{ op } y) = (x \text{ op } y)(1 + \delta), \qquad |\delta| \le u, \quad \text{op} = *, /$$

$$A = J_3(0), \quad B = J_3(10^{-3}), \quad c_{ij} \equiv 1$$
$$\eta(y) = \min\{\epsilon : (A + E)y = b + f, \ \|E\| \le \epsilon\|A\|, \ \|f\| \le \epsilon\|b\|\}.$$

These displays were typed in LaTeX as follows (for display here I have wrapped some of the lines so that they fit comfortably in the given page width).

```
\def\op{\mathbin{\rm op}}
\begin{eqnarray*}
       \|A^T\|_p &=& \|A\|_q, \qquad \frac{1}{p} +
                                    \frac{1}{q} = 1 \\
   (A+E)\widehat x &=& b, \qquad \|E\| \le \epsilon \|A\| \\
      fl(x \op y) &=& (x \op y)(1+\delta),
                          \qquad |\delta|\le u,
                          \quad \op = *,/
\end{eqnarray*}
\begin{eqnarray*}
   & A = J_3(0), \quad B = J_3(10^{-3}),
                  \quad c_{ij} \equiv 1 & \\
   & \eta(y) = \min \{\, \epsilon: (A+E)y = b+f, \;\;
                       \|E\| \le \epsilon \|A\|, \;\;
                       \|f\| \le \epsilon \|b\| \,\}. &
\end{eqnarray*}
```

A \quad is also useful for separating words from symbols:

$$\text{minimize} \quad F(x) \quad \text{subject to} \quad c(x) = 0$$
$$y'(t) = F(t, y(t), y(t - \tau)), \quad y(t) = \Phi(t) \text{ for } t \in [-\tau, 0]$$
$$g(n) \ge \frac{n+1}{2} \quad \text{for all } n.$$

These displays were typed in LATEX as follows.

```
\begin{eqnarray*}
   & \mbox{minimize} \quad F(x) \quad \mbox{subject to}
                       \quad c(x)=0 & \\
   &  y'(t) = F(t, y(t), y(t-\tau)),
                   \quad \mbox{$y(t) = \Phi(t)$
                     for $t\in[-\tau,0]$} & \\
   & g(n) \ge \displaystyle\frac{n+1}{2}
                   \quad \mbox{for all $n$}. &
\end{eqnarray*}
```

Delimiters

Be wary of the \left and \right commands for automatically sizing delimiters, as they often produce symbols that are too large. A \big, \Big or \bigg command (in order of increasing size) usually gives the appropriate size. Here are three examples.

$$c(A) = \min \left\{ \epsilon : \sum_{i=0}^{\infty} |G^i M^{-1}| \le \epsilon |(I - G)^D M^{-1}| \right\}, \quad (\text{\left})$$

$$c(A) = \min\left\{ \epsilon : \sum_{i=0}^{\infty} |G^i M^{-1}| \le \epsilon |(I - G)^D M^{-1}| \right\}, \quad (\texttt{\textbackslash Big})$$

$$\|S_m\|_2 = \left(3 + \frac{1}{1 - \lambda_0} \right) \xi_N, \quad (\texttt{\textbackslash left})$$

$$\|S_m\|_2 = \left(3 + \frac{1}{1 - \lambda_0} \right) \xi_N, \quad (\texttt{\textbackslash Big})$$

$$G\left(x^{(k)} \right) = -g\left(x^{(k)} \right), \quad (\texttt{\textbackslash left}) \qquad G\left(x^{(k)} \right) = -g\left(x^{(k)} \right) \quad (\texttt{\textbackslash big}).$$

Note that \big does not work in LaTeX 2.09 unless the 10pt style and \normalsize are selected. A cure is available, for example, in a style file fixup.sty by John Hobby (details of this and other style files can be found in TeX-index; see page 172).

Miscellaneous Mathematics

The colon character is treated as a relational operator in mathematics mode, as in $x := y$ and $\{ z : |z| \le 1 \}$. If you want a colon to be treated as a punctuation mark use \colon. Thus, $A(1{:}r, 1{:}r)$ is typed as $A(1\colon r,1\colon r)$.

Recall that mathematical functions should be typeset in roman (see page 27). This rule is frequently violated by TeX users, giving aesthetically disturbing results such as $sign(det(A)) = -1$ instead of $\text{sign}(\det(A)) = -1$. If you are using a function for which there is not a predefined macro, define one of the following form

    `\def\diag{\mathop{\rm diag}}`

Then type formulas such as $D = \text{diag}(a_{ii})$ as $D = \diag(a_{ii})$.

I prefer to use the accents \widehat and \widetilde ($\widehat{x}$, $\widetilde{x}$) instead of \hat and \tilde ($\hat{x}$, $\tilde{x}$) because the wide versions are easier to read and are less likely to be mistaken for a stray blob on a photocopied page.

Exercise restraint in the use of TeX's mathematical formatting within the text—elaborate formulas can look out of place and tall ones can affect the line spacing. A common mistake is to use \frac instead of a slash: $x = (y + z)/2$ looks better than $x = \frac{y+z}{2}$, and $\frac{a}{b}$ is better written as ab^{-1} or a/b.

An ellipsis (the sequence of three dots "...") is used to mark omission from text or in a mathematical formula such as "$i = 1, 2, \ldots, n$". In both cases if you type the ellipsis as three full stops in TeX the spacing is too tight: $i = 1, 2, ..., n$. Instead, the control sequence \dots should be used (or \cdots for centred dots in a formula—see page 27). Vertical and diagonal ellipses for use in matrices are obtained with

\vdots (:) and \ddots ($\cdot\cdot$).

Figures in TEX

Figures can be drawn in LATEX using the `picture` environment, which has a repertoire of circles, ovals, and straight lines at a limited number (48) of slopes. (Figure 10.2, below, was drawn this way.) Another set of macros, PICTEX by Michael Wichura, can be used to produce pictures in TEX and LATEX. PICTEX is freely available and the manual can be purchased from the TEX Users Group. Because of the limitations of TEX for drawing pictures it is often necessary to produce figures with a program designed specifically for the purpose. This is most commonly done using figures represented in PostScript.[2]

A widely used set of macros for TEX and LATEX called `psfig` allows PostScript figures to be included with a command of the form

`\psfig{psfile=myfig.ps}`

This command leaves the appropriate amount of vertical space for the figure, and, assuming the macros have been properly customized for the given dvi to PostScript driver, inserts a `\special` command in the dvi file that causes the figure to be included. (The TEX `\special` command is transmitted unchanged to the dvi file and was designed as a means for giving instructions to a dvi driver, for example, to include a figure or to change colour on the printer.) Various options allow the figure to be scaled. Another set of macros called `epsf` works similarly; these were written by Tomas Rokicki of Radical Eye Software for use with his popular `dvips` driver.

Gnuplot[3] is a widely available command-driven program for drawing graphs and plotting data. It displays its results on the screen and can also render its output in the form of a PostScript file or a LATEX `picture` environment. In the latter case hundreds or thousands of plot commands may be needed and you may find that your LATEX program runs out of memory when it processes the file. Figure 10.1 illustrates Gnuplot's LATEX output (the data is concocted, so I have left the axes unlabelled).

---

[2]PostScript is a page description language of Adobe Systems, Inc. A PostScript file is an ASCII file of PostScript language commands that specifies how the output is to appear on the page. By convention, the first line of a PostScript file begins with the two characters %!. Ghostscript is an interpreter for PostScript that is useful for printing PostScript files on non-PostScript printers and previewing them on the screen. It was written by Aladdin Enterprises, Palo Alto, and runs on various Unix machines and under MS-DOS.

[3]Written by Thomas Williams and Colin Kelley at Villanova University in Pennsylvania.

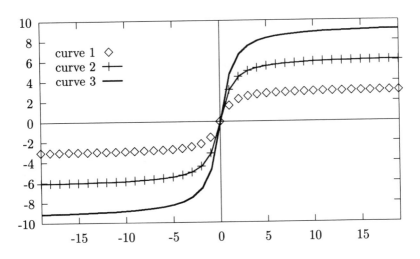

Figure 10.1. Gnuplot output.

10.3. BIBTEX

BIBTEX is a valuable aid to preparing reference lists with LATEX. It was written by Oren Patashnik and is discussed in [131, App. B] and [162]. (Similar tools called bib and refer exist for use with troff.) BIBTEX is a program that reads a LATEX aux (auxiliary) file and constructs a sorted reference list, making use of one or more **bib** files that comprise a database of papers in the BIBTEX format. This reference list (contained in a **bbl** file) is read and processed by LATEX. A diagram showing how LATEX interacts with BIBTEX and MakeIndex is given in Figure 10.2 (MakeIndex is described in the next section).

A **bib** file is an ASCII file that is maintained by the user (in the same way as a **tex** file). As an example, references [122] and [124] are expressed in BIBTEX format as

```
@article{knut79,
    author = "Donald E. Knuth",
    title = "Mathematical Typography",
    journal = "Bull. Amer. Math. Soc. (New Series)",
    volume = 1,
```

```
    number = 2,
    pages = "337-372",
    year = 1979
      }

@book{knut86,
    author = "Donald E. Knuth",
    title = "The {\TeX book}",
    publisher = "Addison-Wesley",
    address = "Reading, Massachusetts",
    year = 1986
      }
```

The first part of the last sentence was typed as

```
As an example, references \cite{knut79} and \cite{knut86} are
expressed in \BibTeX\ format as
```

There are three advantages to using BIBTEX.

1. By specifying the appropriate bibliography style (**bst** file) from the several available it is trivial to alter the way in which the references are formatted. Possibilities include BIBTEX's **plain**, **abbrv** and **alpha** formats, illustrated, respectively, by

> [1] Donald E. Knuth. Mathematical typography. *Bulletin Amer. Math. Soc. (New Series)*, 1(2):337–372, 1979.
> [1] D. E. Knuth. Mathematical typography. *Bulletin Amer. Math. Soc. (New Series)*, 1(2):337–372, 1979.
> [Knu79] Donald E. Knuth. Mathematical typography. *Bulletin Amer. Math. Soc. (New Series)*, 1(2):337–372, 1979.

With the **alpha** format, citations in the text use the alphanumeric label constructed by BIBTEX ([Knu79] in this example). The **unsrt** format is the same as **plain** except it lists the entries in order of first citation instead of alphabetical order (as is required, for example, by the journal *Computers and Mathematics with Applications*). In this book I am using the **siam** format, which formats references in the style used by SIAM journals; a distinctive feature of this format is the use of the 3-em dash for repeated author names—see the entry for [122], for example.

2. To keep the reference lists of working papers up to date (as technical reports become journal papers, for example) it is necessary only to

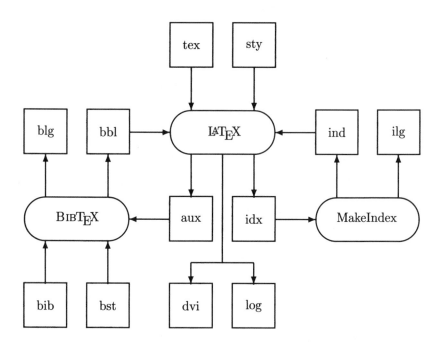

Figure 10.2. Interaction between LaTeX, BibTeX and MakeIndex. The `log`, `blg` and `ilg` files contain error messages and statistics summarizing the run. The `aux` file contains cross-referencing information. The `sty` and `bst` files determine the style in which the LaTeX document and BibTeX reference list are produced.

update the master **bib** file and rerun LATEX and BIBTEX on each
paper. If BIBTEX were not used, the reference list in each paper
would have to be updated manually.

3. Since BIBTEX is widely used and available on a wide range of machines
   it is possible for people to exchange and share databases. In the
   Numerical Analysis community the BIBTEX databases for the book by
   Golub and Van Loan [84] (861 entries) and a published bibliography
   on parallel computing [155] (2016 entries) are available by electronic
   mail from netlib [54] (see §10.9). The database for Björck's survey
   [24] (500+ entries) is available by anonymous ftp, as is the personal
   database of numerical analyst G. W. Stewart (2000+ entries) [196].

   Some research groups maintain a common BIBTEX database accessible
   to all users on their computer system, thereby rapidly building up a
   large database with minimal duplication of effort.

Aside from these BIBTEX-specific reasons, there are other reasons for
keeping a personalized computer database of references. If you record every
paper you read then a computer search allows you to check whether or not
you have read a given paper. This is a very handy capability to have, as once
you have read more than a couple of hundred papers it becomes difficult
to remember their titles and authors. Also, you can put comments in the
database to summarize your thoughts about papers. Some BIBTEX users
include an **annotate** field in their **bib** files. Although standard BIBTEX
style files ignore this field, other style files are available that reproduce it,
and they can be used to prepare annotated bibliographies. In my BIBTEX
book entries I include an **isbn** field. To prepare the bibliography for this
book I modified the **siam.bst** file so that it puts the ISBN entry at the end
of a reference.

It is important to note that if names are typed in a **bib** database with
no space between the initials (e.g., **D.E. Knuth**), BIBTEX produces only
the first initial. You should therefore always leave a space between initials
(**D. E. Knuth**), as is standard practice in typesetting. Even better is to
type the full name, if it is given.

A minor disadvantage of BIBTEX is that there are no standard utilities
to help maintain the database. It is up to the user to keep the database
in alphabetical order (if desired) and to format the entries in a readable
way. One way to sort a database is with a short AWK program. AWK
is an interpreted programming language available on most Unix systems
[4]. A program for sorting multiline records given in §3.4 of [4] is easily
adapted to sort BIBTEX databases. It collapses multiline records onto a
single line, sorts using the Unix **sort** command, then transforms back to

the original multiline format. AWK is also handy for searching a `bib` file. The command line

```
awk 'BEGIN{ RS="" } /Riemann/' my.bib
```

searches the file `my.bib` in the current directory and prints all the entries that contain the word Riemann (assuming that records are separated by a blank line). Note that this AWK call prints complete `bib` entries, not just the lines on which the word occurs, as a `grep` search would.

Sorting and searching of `bib` files can also be done using public domain tools made available by their authors over the computer networks. One particularly useful tool is `bibclean` by Nelson H. F. Beebe, which "pretty-prints" a `bib` file and checks for syntax errors; it even checks ISBN and ISSN fields to see whether the checksum is correct.

In my experience, the flexibility of BIBTEX makes it well worth using. While it might appear tedious to enter references in the required format, this can be done quickly with practice, by either filling in a blank template or editing an existing reference.

10.4. Indexing

The purpose and content of a book index are well explained in *The Chicago Manual of Style* [212]:

> Every serious book of nonfiction should have an index if it is to achieve its maximum usefulness. A good index records every pertinent statement made within the body of the text.... The subject matter and purpose of the book determine which statements are pertinent and which peripheral. An index should be considerably more than an expanded table of contents and considerably less than a concordance of words and phrases.

The preparation of an index requires an appreciation of what makes a good index, together with skill and hard work. A good index helps the reader to pinpoint a topic of interest or to determine that the topic is not covered in the book. It contains surprises—pointers to passages that the reader might overlook when scanning the book and its table of contents. It anticipates the various ways in which a reader might search for a topic, by including it under multiple entries, where appropriate. For example, *block LU factorization* might be listed under *block, factorization, LU,* or even *decomposition* or *matrix.* One source of index entries is section and subsection headings, since these provide the framework of the text. Entries should be nouns or noun phrases, not adjectives.

Index preparation is the aspect of book production that is least often computerized, partly because it tends to be regarded as a one-off task to be done once a book is complete (as it must be in traditional typesetting). However, indexing is just as amenable to computer help as other aspects of document preparation.

MakeIndex is a C program, written by Pehong Chen [42], [133] with advice from Leslie Lamport, that makes an index for a LaTeX document. The user has to place \index commands in the LaTeX source that define the name and location of the items to be indexed. If a \makeindex command is placed in the preamble (before \begin{document}) then LaTeX writes the index entries, together with the page numbers on which they occur, to an idx file. This is read by the MakeIndex program, which processes and sorts the information, producing an ind file that generates the index when included in the LaTeX document. MakeIndex supports various options in the \index command to support standard indexing requirements, such as subentries, page ranges, and cross references to other entries. Here is how the beginning of one sentence from page 152 was typed:

```
\item  It is easy to prepare
transparencies\index{slides!preparing in \TeX}
with \TeX\ if the paper
```

The exclamation mark in the \index command denotes the beginning of a subentry.

AWK tools for indexing are described in [4, §5.3] and [18]; these tools do not support subentries. A simple and elegant way to construct key word in context (KWIC) indexes using AWK is also described in [4]. A KWIC index lists each word in the context of the line in which it is found; the list is sorted by word and arranged so that the key words line up. One of the main uses of KWIC indexes is to index titles of papers.

The best advice I have seen on preparing an index is that given in *The Chicago Manual of Style*. Other good references are *Words Into Type* [184] and *Copy-Editing* [33]. Most published advice, though, was written before the advent of computer indexing and describes manual techniques involving 3 × 5 index cards. If you are indexing with LaTeX and MakeIndex it is advisable *not* to leave the task to the very last stage. For, in inserting the \index entries, you are likely to introduce errors (of spacing, at least) and so a further round of proofreading will be needed after the indexing stage.

For an interesting example of an index, see Halmos's *I Want to Be a Mathematician* [97]. Priestley [173] says "If index writing has not bloomed into an art form in *I Want to Be a Mathematician*, it has at least taken a quantum leap forward. From 'academic titles, call me mister' to 'Zygmund, A., at faculty meetings' this one is actually worth reading."

10.5. Text Editors

As the use of computers in research and writing increases, we spend more and more time at the keyboard, much of that spent typing text with a text editor. Of all the various programs we use, the text editor is the one that generates the most extreme feelings: most people have a favourite editor and fervently defend it against criticism. Under the Unix operating system two editors are by far the most widely used. The first, and oldest, is vi, which has the advantage that it is available on every Unix system. The second, and the more powerful, is Emacs. Not only does Emacs do almost everything you would expect of a text editor, but from within it you can run other programs and view and edit their output; rename, move and delete files; and send and read electronic mail. Some workstation users carry out nearly all their computing "within Emacs", leaving it running all the time.

There are various versions of Emacs, one of the most popular of which is GNU Emacs [38], [190]. (GNU stands for "Gnu's not UNIX" and refers to a Unix-like operating system that is being built by Richard Stallman and his associates at the Free Software Foundation.) GNU Emacs is available for workstations and 386-based PC-compatibles; other PC versions include Freemacs, MicroEmacs and Epsilon. GNU Emacs contains modes for editing special types of files, such as TEX, LATEX and Fortran files. In these modes special commands are available; for example, in TEX mode one Emacs command will invoke TEX on the file in the current editing buffer. Appendix C contains a list of the 60+ most useful GNU Emacs commands and should be helpful to beginners and intermediate users.

10.6. Spelling Checking, Filters and Pipes

Programs that check, and possibly correct, spelling errors are widely available. These are useful tools, even for the writer who spells well, because, as McIlroy has observed [146], most spelling errors are caused by errors in typing. The Unix[4] spelling checker spell takes as input a text file and produces as output a list of possibly misspelled words. The list comprises words that are not in spell's dictionary and which cannot be generated from its entries by adding certain inflections, prefixes or suffixes; a special

---

[4]More details on the Unix commands described in this section can be obtained from a Unix reference manual or from the on-line manual pages by typing man followed by the command name.

"stop list" avoids non-words such as *beginer* (*begin* + *er*) being accepted. The development of `spell` is described in a fascinating article by McIlroy [146] and is summarized by Bentley in [16, Chap. 13]. When I ran an earlier version of this book through `delatex` (see below), followed by `spell` with the British spelling option, part of the output I obtained was as follows (the output is wrapped into columns here to save space)

bbl	capitalized	de	diag
beginer	cccc	deci	dispensible
blah	co	delatex	dvi
blocksize	comp	dependant	ees
bst	computerized	der	eg
capitalization	crr	detex	eigenproblem

The -ize endings are flagged as errors because `spell` expects you to use -ise endings when the British spelling option is in effect. The one genuine error revealed by this extract is *dispensible*, which should be *dispensable*. `Spell` can be instructed to remove from its output words found in a supplemental list provided by the user. This way you can force `spell` to accept technical terms, acronyms and so on. `Spell` can be used to check a single word, *mathematical*, say, by typing at the command line

```
spell
mathematical
^d
```

(`^d` is obtained by holding down ctrl and typing d). In this case there is no output because the word is recognized by `spell`.

If you use GNU Emacs, you can call the Unix `spell` program from within the editor. The command `Esc-$` checks the word under the cursor and `Esc-x spell-buffer` checks the spelling of the whole buffer. In each case you are given the opportunity to edit an unrecognized word and replace all occurrences with the corrected version.

A problem with spell checking TEX documents is that most TEX commands will be flagged as errors. When working in Unix the solution is to run the file through `detex`, or `delatex`[5] for LATEX documents, before passing it to `spell`; these filters strip the file of all TEX and LATEX commands, respectively. An alternative is simply to have the spelling checker learn the TEX and LATEX command names as though they were valid words.

It is important to realize that spelling checkers will not identify misspellings that are themselves words, such as *form* for *from* (a common

---

[5]This filter is available from netlib in the typesetting directory (see §10.9).

error in published papers), *except* for *expect*, or *conversation* for *conversation*. For this reason, bigger does not necessarily mean better for dictionaries used by spelling checkers. Peterson [168] investigates the relationship between dictionary size and the probability of undetected typing errors; he recommends that "word lists used in spelling programs should be kept small."

Spelling correction programs are also available on various computers. These not only flag unrecognized words, but present guesses of the correct spelling. They look for errors such as transposition of letters, or a single letter extra, missing or incorrect (research is mentioned by Peterson in [167] that finds these to be the cause of 80% of all spelling errors). The suggested corrections can be amusing: for example, one spelling corrector I have used suggests dunce for Duncan and turkey for Tukey.

Ispell is an interactive spelling checker and corrector available for Unix and DOS systems. When invoked with a filename it displays each word in the file that does not appear in its dictionary and offers a list of "near misses" and guesses of the correct word. You can accept one of the suggested words or type your own replacement. An Ispell interface exists for GNU Emacs.

You can do limited searching of **spell**'s dictionary using the **look** command, which finds all words with a specified prefix. Thus

```
look comp | grep ion$
```

displays all words that begin with *comp-* and end in *-ion*.

It is interesting to examine the frequency of word usage in your writing. Under Unix this can be done with the following pipe, where **file** is a filename:

```
cat file | deroff -w | tr A-Z a-z | sort | uniq -c | sort -rn
```

The **deroff -w** filter divides the text into words, one per line (at the same time removing any troff commands that may be present), **tr A-Z a-z** converts all words to lower case, **sort** sorts the list, **uniq -c** converts repeated lines into a single line preceded by a count of how many times the line occurred, and **sort -rn** sorts on the numeric count field in reverse order (largest to smallest). Applying this pipe to an earlier version of this book I obtained as the first twenty-five lines of output (wrapped into columns here to save space)

1533 the	529 and	266 for	185 by	154 i
709 a	517 in	259 that	175 as	154 you
696 of	310 ndex	230 are	167 egin	150 not
690 to	275 it	220 mph	164 nd	146 or
613 is	267 be	186 ite	160 this	141 with

(The non-words **ndex**, **mph**, **ite**, **egin**, and **nd** are left-over LaTeX commands, which appear since I did not use **delatex** in the pipe.) It is worth examining word frequency counts to see if you are overusing any words. As far as I am aware, this particular count does not reveal any abnormalities in my word usage.

The Unix **diff** command takes two files and produces a list of changes that would have to be made to the first file to make it the same as the second. The changes are expressed in a syntax similar to that used in the **ed** text editor. If you use the **-c** option (**diff -c filename**) then three lines before and after each change are printed to show the context. The main use of **diff** for the writer is to see how two versions of a file (a current and an earlier draft, say) differ. If your co-author updates the source file for a TeX paper, you can use **diff** to see what changes have been made.

Another useful filter is the **wc** command, which counts the lines, words and characters in a file. When I ran the source for an almost-final draft of this book through **wc** (using the command **wc *.tex**, since the source is contained in several **.tex** files) I obtained the output

```
10900    59873   413716    total
```

This count shows that the source contains 59,873 words, though many of these words are LaTeX instructions that do not result in a printed word, so this is an overestimate of the actual word count. When I ran the source through **delatex** before sending it to **wc** the word count dropped to 49,191.

On some computer systems on-line thesauri are available that, at the press of a key, look up the word under the cursor and present a list of synonyms. A computer-readable version of the 1911 edition of Roget's Thesaurus is available in the public domain from Project Gutenberg (for information send electronic mail to Michael Hart (**hart@vmd.cso.uiuc.edu**)).

10.7. Style Checkers

Programs exist that try to check the style of your text. Various commercial programs are available for PCs. Style checking programs have been available for Unix machines since the late 1970s. An article by Cherry [43] describes several programs: these include **style**, which "reads a document and prints a summary of readability indices, sentence length and type, word usage, and sentence openers," and **diction**, which "prints all sentences in a document containing phrases that are either frequently misused or indicate wordiness."

One of the readability formulas used by **style** is the Kincaid formula, which assigns the reading grade (relative to the US schooling system)

11.8 × (average syllables per word) + 0.39 × (average words per sentence) − 15.59.

The formula was derived by a process that involved measuring how well a large sample of US Navy personnel understood Navy technical manuals. (It is a contractual requirement of the US Department of Defense that technical manuals achieve a particular reading measure.)

In his book *The Art of Plain Talk* [61, Chap. 7], Flesch proposes the following score for measuring the difficulty of a piece of writing:

$$s = 0.1338 \times \text{(average words per sentence)}$$
$$+ 0.0645 \times \text{(average affixes per 100 words)}$$
$$- 0.0659 \times \text{(average personal references per 100 words)} - 0.75.$$

The score is usually between 0 and 7, and Flesch classifies the scores in unit intervals from $s \leq 1$, "very easy", to $s > 6$, "very difficult". He states that comics fall into the "very easy" class and scientific journals into the "very difficult" class. Klare [120] explains that "Prior to Flesch's time, readability was a little-used word except in educational circles, but he made it an important concept in most areas of mass communication."

The limitations of readability indices are well known and are recognized by their inventors [120], [145]. For example, the Kincaid and Flesch formulas are invariant under permutations of the words of a sentence. More generally, readability formulas measure style and not clarity, content or organization. For the writer, a readability formula is best regarded as a rough means for rating a draft to see whether it is suitable for the intended audience.

AT&T's Bell Laboratories markets *The Writer's Workbench*, an extensive system that incorporates **style**, **diction** and various other programs [139]. One of these is **double**, which checks for occurrences of a word twice in succession, possibly on different lines. Repeated words are hard for a human proofreader to detect.

Hartley [101] obtained suggestions from nine colleagues about how to improve a draft of his paper [100] and compared them with the suggestions generated by *The Writer's Workbench*. He concluded that "Text-editing programs can deal well with textual issues (perhaps better than humans) but humans have prior knowledge and expertise about content which programs currently lack."

Knuth ran a variety of sample texts through **diction** and **style** [128, §40], including technical writing, *Wuthering Heights* and Grimm's Fairy

Tales. He found that his book of commentaries on Chapter 3, verse 16 of each book of the Bible [126] was given a significantly lower reading grade level than the other samples, and concluded that we tend to write more simply when writing outside our own field.

10.8. Review, Abstract and Citation Services

When you need to find out what work has been done in a particular area, or need to track down an incomplete reference, you should consult one of the reviews or citation collections. The main ones are as follows.

Mathematical Reviews (MR) is run by the American Mathematical Society (AMS) and was first published in 1940. Each month the journal publishes short reviews of recently published papers drawn from approximately 2000 scholarly publications. Each review either is written by one of the nearly 12,000 reviewers or is a reprint of the paper's abstract. The reviews are arranged in accordance with the Mathematics Subject Classifications (see §6.7). MR is particularly useful for finding details of a paper in a journal that your library does not receive—based on the review you can decide whether to order the paper via the inter-library service. Sometimes you will see an entry in a reference list containing a term such as MR 31 #1635. This means that the article in question was reviewed in volume 31 of MR as review number 1635.

The AMS also produces *Current Mathematical Publications*, which is essentially a version of MR that contains the bibliographic records but not the reviews. However, CMP is much more up to date than MR: it is issued every three weeks and contains a list of items received by the MR office, most of which will eventually be reviewed in MR.

Computing Reviews plays a role for computer science similar to the one MR plays for mathematics. It is published by the Association for Computing Machinery and uses its own classification scheme (see §6.7). The other major abstracting journal for computer science is *Computer and Control Abstracts*; it has wider coverage than CR.

Current Contents (from the Institute for Scientific Information (ISI), Philadelphia) is a weekly list of journal contents pages (similar in size to the US *TV Guide*). The Physical Sciences edition is the one in which mathematics and computer science journals appear. Each issue of CC is arranged by subject area and contains an index of title words. CC is also available on diskette, optionally with abstracts. Each edition of CC contains an article by Eugene Garfield, the founder of ISI. These articles often report citation statistics, such as most-cited papers in particular subject areas.

The *Science Citation Index* (SCI), also from the ISI, records all papers that reference an earlier paper, across all science subjects. It began in the early 1960s and covers the period from 1945 to the present [67], [70]. The SCI provides a means for finding newer papers that were influenced by older ones, whereas searching reference lists takes you in the opposite direction. For example, suppose we are interested in the reference

> W. KAHAN, *Further remarks on reducing truncation errors*, Comm. ACM, 8 (1965), p. 40.

If we look under "KAHAN W" in the five-year cumulation 1975–1979 *Citation Index*, and then under the entry for his 1965 paper, we find four papers that include Kahan's in their reference lists. For each of these citing papers the first author, journal, volume, starting page number and year of publication are given; the years range from 1975–1977. The full bibliographic data for these papers can be found in the SCI *Source Index*. Looking up these four papers in later indexes we find further references on the topic of interest.

If you know a key author in an area, or can remember the title of a paper but not the author, the SCI *Permuterm Subject Index* can help. This is a key word index in which every significant word in each article title is paired with every other significant word from the same title. Under each pair of key words is a list of relevant authors; their papers may be found in the *Source Index*. As an example, if all we know of Kahan's article are the words "truncation" and "errors" and the year of publication we can find the full details of the article from the five-year cumulation 1965–1969 *Permuterm Subject Index* and the corresponding *Source Index*. Garfield's article "How to Use the Science Citation Index" [78] gives detailed examples of the use of the SCI and is well worth reading.

Zentralblatt für Mathematik und ihre Grenzgebiete (ZM), also titled *Mathematical Abstracts*, is another mathematical review journal. It was founded in 1931 and is published by Springer-Verlag and Fachinformationszentrum Karlsruhe. It uses the Mathematics Subject Classifications, and its coverage is almost identical to that of MR.

MR, SCI and ZM are available in computer-readable form on compact disc (CD-ROM) and in on-line databases.

10.9. Internet Resources

In this section I describe some of the computer facilities available over the Internet network that are of general interest to mathematicians. Note that services of this type are usually run on a voluntary basis. There is no guarantee that a particular service will continue and the details of how to access it may change. For a thorough explanation of the Internet and

Citation Facts and Figures

The *Science Citation Index* (1945–1988) contains about 33 million cited items. The most-cited paper is

O. H. LOWRY, N. J. ROSEBROUGH, A. L. FARR AND
R. J. RANDALL, *Protein measurement with the Folin phenol reagent*, J. Biol. Chem., 193 (1951), pp. 265–275,

which has 187,652 citations. The next most cited paper (also on protein methods, as are all of the top three) has 59,759 citations. The six most-cited papers from Mathematics, Statistics and Computer Science are as follows; their rankings on the list of most-cited papers range from 24th to 297th.

1. D. B. DUNCAN, *Multiple range and multiple F tests*, Biometrics, 11 (1955), pp. 1–42. (8985 citations)

2. E. L. KAPLAN AND P. MEIER, *Nonparametric estimation from incomplete observations*, J. Amer. Statist. Assoc., 53 (1958), pp. 457–481. (4756 citations)

3. D. W. MARQUARDT, *An algorithm for least-squares estimation of nonlinear parameters*, J. Soc. Indust. Appl. Math., 11 (1963), pp. 431–441. (3441 citations)

4. D. R. COX, *Regression models and life-tables*, J. Royal Statist. Soc. Ser. B Metho., 34 (1972), pp. 187–220. (3392 citations)

5. R. FLETCHER AND M. J. D. POWELL, *A rapidly convergent descent method for minimization*, Comp. J., 6 (1963), pp. 163–168. (1948 citations)

6. J. W. COOLEY AND J. W. TUKEY, *An algorithm for the machine calculation of complex Fourier series*, Math. Comp., 19 (1965), pp. 297–301. (1845 citations)

In the period 1945–1988, 55.8% of the papers in the index were cited only once, and 24.1% were cited 2–4 times. 5767 papers were cited 500 or more times. References: [72], [73], [75], [76], [77].

its resources I recommend *The Whole Internet User's Guide & Catalog* by Krol [130]. As well as explaining electronic mail, telnet, ftp, and network news, this book describes the use of tools such as Archie, gopher, Wide Area Information Servers (WAIS), and the World-Wide Web (WWW). For a less detailed description of these topics see *The Internet Companion* by LaQuey and Ryer [134].

Newsgroups

The news system available on many computer networks contains a large number of newsgroups to which users contribute messages. The newsgroups frequently carry announcements of new software and software updates. On a Unix system, type `man rn` or `man nn` for information on how to read news. Newsgroups of general interest to mathematicians are `sci.math` and its more specialized cousins, such as `sci.math.research` and `sci.math.symbolic`, and `comp.text.tex` for TEX information. For many newsgroups a FAQ document of Frequently Asked Questions is available. Of particular interest to TEX users are `tex-faq`, maintained by Bobby Bodenheimer, and an STI (Supplementary TEX Information) file maintained by Guoying Chen, both available on `comp.text.tex`. Another useful file is `TeX-index`, an index maintained by David M. Jones of all TEX macros that are available via anonymous ftp or mail-servers. An example of an entry in `TeX-index` is the following one for `siam.sty`, which is a LATEX style file used in typesetting SIAM papers.

```
Name: siam.sty
Description: LaTeX document style for producing articles for
    SIAM journals.
Keywords: LaTeX, article, journal, publisher style
Author: Howard Trickey <trickey@research.att.com>
Supported: yes
Latest Version: 5 Nov 1991
Archives: research.att.com* (in netlib/typesetting
    /siamltex.all), ymir
See also: amstexsiam.sty, siam10.sty, siam11.sty, siam12.sty,
    siam.bst, fixup.sty, numinsec.sty
```

Digests

Various magazines are available by electronic mail. These collect questions, answers and announcements submitted to a particular electronic mail address. For example, NA-Digest is a weekly magazine about numerical analysis [55]; send mail to `na.help@na-net.ornl.gov` for information on how to subscribe. T_EXhax and UKT_EX Digest are concerned with T_EX and are distributed when a sufficiently large issue has built up. To subscribe, send a message containing the word "subscribe" and your email address to `texhax@tex.ac.uk` and `uktex-request@tex.ac.uk`, respectively. Both of these digests also appear in the `comp.text.tex` newsgroup.

Netlib

Netlib is an electronic repository of public domain mathematical software for the scientific computing community. In particular, it houses the various -PACK program libraries, such as EISPACK, LINPACK, MINPACK and LAPACK, and the collected algorithms of the Association for Computing Machinery (ACM). Netlib has been in existence since 1985 and can be accessed by electronic mail, by ftp, or through an interactive X windows interface called xnetlib. In addition to providing mathematical software, netlib provides the facility to download technical reports from certain institutions, to download software and errata for textbooks, and to search the SIAM membership list (via the `whois` command). Background on netlib is given in an article by Dongarra and Grosse [54] and news of the system is published regularly in NA-Digest and *SIAM News*. To obtain a catalogue of the contents of netlib send an email message with body **send index** to `netlib@ornl.gov` or `netlib@research.att.com`. Copies of netlib exist at various other sites throughout the world.

e-MATH

The AMS runs a computer service, e-MATH, with many features. For example, it allows you to obtain pointers to reviews in *Mathematical Reviews* (1985 to present), to download the list of Mathematics Subject Classifications, to download articles (in A_MS-T_EX format) from the *Bulletin of the American Mathematical Society* journal, and to access lists of employment and post-doctoral opportunities. At the time of writing, e-MATH can be accessed over the Internet by `telnet` to the address `e-math.ams.org` (or its numeric form, `130.44.1.100`). The login and password are both `e-math`. Status reports on e-MATH are published in the *Notices of the American Mathematical Society* in the "Inside the AMS" column (every personal and institutional member of the AMS receives this journal).

10.10. Anonymous Ftp

Ftp stands for file transfer protocol, and is a standard means for interactively transferring files between different computers (particularly Unix machines) on the Internet network. Normally, to transfer a file from one computer to another by ftp, you have to log onto one machine from the other via the ftp program, and so you must have an account on both machines. Anonymous ftp is a special form of ftp in which you log on as user **anonymous** and need not type a password (though, by convention, you are supposed to type your electronic mail address to indicate who you are). Anyone can log onto an anonymous ftp account and transfer files from it— all they need to know is the account name and address. Thus anonymous ftp is a convenient way to make papers (in the form of TeX dvi files, or PostScript files), bibliographies, programs, etc., available to others.

Here is an example of an anonymous ftp session in which I log from a Sun workstation at Manchester into a netlib server at AT&T Bell Laboratories, and download the file named `siamltex.all.Z` (a compressed shar file containing the SIAM macros for typesetting papers in LaTeX). The commands that I type appear after the % and `ftp>` prompts. Note that the password is not shown in this example, as it is not echoed to the screen when you type it. The output here is slightly rearranged to fit on the page.

```
vortex 1% ftp research.att.com
Connected to research.att.com.
220 inet FTP server (Version 4.266 Wed Jul 17 14:15:38 EDT
                     1991) ready.
Name (research.att.com:higham): anonymous
331 Guest login ok, send ident as password.
Password:
230 Guest login ok, access restrictions apply.
ftp> cd netlib
250 CWD command successful.
ftp> cd typesetting
250 CWD command successful.
ftp> dir
200 PORT command successful.
150 Opening ASCII mode data connection for /bin/ls.
-rw-rw-r--  1  106  bin  15777 Aug 24  1988  chapman.Z
-rw-rw-r--  1  106  bin  13116 Dec  3  1987  chem.Z
-rw-rw-r--  1  106  bin  23344 Jan 30  1987  cmd-latex.Z
-rw-rw-r--  1  106  bin  19461 Feb  3  1990  imappt-amstex.Z
-rw-rw-r--  1  106  bin   1205 Mar 25  1992  index
-rw-r--r--  1  106  bin   6533 Aug 24  1988  indexing.tools.Z
-rw-rw-r--  1  106  bin  20238 Mar 25  1992  ltexfrnt.all.Z
-rw-r--r--  1  106  bin  13259 Oct  5 16:02  ltexpprt.all.Z
-rw-rw-r--  1  106  bin  13597 Oct  3  1990  ptexpprt.all.Z
```

```
-rw-rw-r-- 1 106  bin   3782 Feb 23  1990 sampltex.all.Z
-rw-rw-r-- 1 106  bin  20012 Feb  3  1990 siamams.all.Z
-rw-rw-r-- 1 106  bin  21895 Nov 11  1991 siamltex.all.Z
-rw-rw-r-- 1 106  bin  14743 May  3  1989 siamptex.all.Z
-rw-rw-r-- 1 106  bin   5517 May 21  1987 siamtrof.all.Z
drwxrwxr-x 2 106  bin    512 Nov 14 11:20 springer
-rw-rw-r-- 1 106  bin   2282 May 24  1992 times.sty.Z
226 Transfer complete.
1008 bytes received in 2 seconds (0.5 Kbytes/s)
ftp> binary
200 Type set to I.
ftp> get siamltex.all.Z
200 PORT command successful.
150 Opening BINARY mode data connection for siamltex.all.Z
    (21895 bytes).
226 Transfer complete.
local: siamltex.all.Z remote: siamltex.all.Z
21895 bytes received in 14 seconds (1.5 Kbytes/s)
ftp> quit
221 Goodbye.
vortex 2%
```

The get command transfers the named file from the remote machine to the local one; the reverse operation is put (though this is usually not allowed in an anonymous ftp session, except in a special contributions directory for making submissions to archives). Both commands take an optional third argument, which is the filename to use on the local machine for get, and the remote machine for put.

When transferring a binary file (such as a TeX dvi file or compressed .Z files) it is important to give the command binary before initiating the transfer. To switch back to the default mode of transfer, type ascii. If you want to download many files use the mget command with a Unix wildcard, as in mget *.dvi. To avoid being prompted for confirmation of each individual transfer, precede the "multiple get" with the prompt command. See the Unix man page (type man ftp) for more details of the ftp command.

There are many anonymous ftp sites in existence. Because of the large number of sites and the possibility that details may change, I do not list any addresses here. To find out about anonymous ftp sites consult newsgroups, as described above.

Table 10.1 lists some of the file types you may encounter when ftp'ing files.

Suffix	Type	Explanation
.bib	ASCII	BibTeX source.
.bst	ASCII	BibTeX style file.
.dvi	binary	TeX output. Use dvips or dvi2ps to convert to PostScript.
.gif, .tif	binary	Image file formats (Graphics Interchange Format, Tagged Image File Format) [28].
.ps	ASCII	PostScript file.
.shar	ASCII	"Shell archive" collection of files. Use sh to extract files.
.sty	ASCII	LaTeX style file.
.tar	binary	"Tape archive" collection of files. Use tar -xvf or pax -r to extract the contents.
.tex	ASCII	TeX source.
.uu	ASCII	Re-coded form of binary file, suitable for mailing. Use uudecode to recover original binary file.
.Z	binary	Compressed. Use uncompress to recover original file.
.z	binary	Compressed by an older algorithm. Use unpack to recover original file.
.zip	binary	Compressed by PKZIP. Use an unzip program to recover original file.

Table 10.1. Standard file types.

Appendix A
The Greek Alphabet

Capital	Lower case	English name
A	α	alpha
B	β	beta
Γ	γ	gamma
Δ	δ	delta
E	ϵ, ε	epsilon
Z	ζ	zeta
H	η	eta
Θ	θ, ϑ	theta
I	ι	iota
K	κ	kappa
Λ	λ	lambda
M	μ	mu
N	ν	nu
Ξ	ξ	xi
O	o	omicron
Π	π	pi
P	ρ, ϱ	rho
Σ	σ, ς	sigma
T	τ	tau
Υ	υ	upsilon
Φ	ϕ, φ	phi
X	χ	chi
Ψ	ψ	psi
Ω	ω	omega

Appendix B
Summary of TₑX and LᴬTₑX Symbols

ò	`\'{o}`	õ	`\~{o}`	ŏ	`\v{o}`	o̧	`\c{o}`	ó	`\'{o}`
ō	`\={o}`	ő	`\H{o}`	o̦	`\d{o}`	ô	`\^{o}`	ȯ	`\.{o}`
o͡o	`\t{oo}`	o̱	`\b{o}`	ö	`\"{o}`	ŏ	`\u{o}`		

Table B.1. Accents.

â	`\hat{a}`	á	`\acute{a}`	ā	`\bar{a}`	ȧ	`\dot{a}`
ă	`\breve{a}`	ǎ	`\check{a}`	à	`\grave{a}`	a⃗	`\vec{a}`
ä	`\ddot{a}`	ã	`\tilde{a}`				

Table B.2. Math mode accents.

œ	`\oe`	Œ	`\OE`	æ	`\AE`
Æ	`\AE`	å	`\aa`	Å	`\AA`
ø	`\o`	Ø	`\O`	ł	`\l`
Ł	`\E`	ß	`\ss`	¿	`?'`
¡	`!'`				

Table B.3. Foreign symbols.

179

α	\alpha	θ	\theta	o	o	τ	\tau
β	\beta	ϑ	\vartheta	π	\pi	υ	\upsilon
γ	\gamma	γ	\gamma	ϖ	\varpi	ϕ	\phi
δ	\delta	κ	\kappa	ρ	\rho	φ	\varphi
ϵ	\epsilon	λ	\lambda	ϱ	\varrho	χ	\chi
ε	\varepsilon	μ	\mu	σ	\sigma	ψ	\psi
ζ	\zeta	ν	\nu	ς	\varsigma	ω	\omega
η	\eta	ξ	\xi				
Γ	\Gamma	Λ	\Lambda	Σ	\Sigma	Ψ	\Psi
Δ	\Delta	Ξ	\Xi	Υ	\Upsilon	Ω	\Omega
Θ	\Theta	Π	\Pi	Φ	\Phi		

Table B.4. Greek letters.

$\pm$	\pm	$\cap$	\cap	$\diamond$	\diamond
$\mp$	\mp	$\cup$	\cup	$\triangle$	\bigtriangleup
$\times$	\times	$\uplus$	\uplus	$\triangledown$	\bigtriangledown
$\div$	\div	$\sqcap$	\sqcap	$\triangleleft$	\triangleleft
$*$	\ast	$\sqcup$	\sqcup	$\triangleright$	\triangleright
$\star$	\star	$\vee$	\vee (or \lor)	$\lhd$	\lhd*
$\circ$	\circ	$\wedge$	\wedge (or \land)	$\rhd$	\rhd*
$\bullet$	\bullet	$\setminus$	\setminus	$\unlhd$	\unlhd*
$\cdot$	\cdot	$\wr$	\wr	$\unrhd$	\unrhd*
$\oplus$	\oplus	$\ominus$	\ominus	$\otimes$	\otimes
$\oslash$	\oslash	$\odot$	\odot	$\bigcirc$	\bigcirc
$\dagger$	\dagger	$\ddagger$	\ddagger	$\amalg$	\amalg
$+$	+	$-$	-		

* predefined in ℐₐTₑX but not in TₑX.

Table B.5. Binary operation symbols.

,	,	;	;	:	\colon	.	\ldotp	$\cdot$	\cdotp

Table B.6. Punctuation symbols.

$\leq$	\leq (or \le)	$\geq$	\geq (or \ge)	$\equiv$	\equiv
$\prec$	\prec	$\succ$	\succ	$\sim$	\sim
$\preceq$	\preceq	$\succeq$	\succeq	$\simeq$	\simeq
$\ll$	\ll	$\gg$	\gg	$\asymp$	\asymp
$\subset$	\subset	$\supset$	\supset	$\approx$	\approx
$\subseteq$	\subseteq	$\supseteq$	\supseteq	$\cong$	\cong
$\sqsubset$	\sqsubset*	$\sqsupset$	\sqsupset*	$\neq$	\neq (or \ne)
$\sqsubseteq$	\sqsubseteq	$\sqsupseteq$	\sqsupseteq	$\doteq$	\doteq
$\in$	\in	$\ni$	\ni (or \owns)	$\notin$	\notin
$\vdash$	\vdash	$\dashv$	\dashv	$<$	<
$\models$	\models	$\perp$	\perp	$\mid$	\mid
$\parallel$	\parallel	$\bowtie$	\bowtie	$\Join$	\Join*
$\smile$	\smile	$\frown$	\frown	$\propto$	\propto
$=$	=	$>$	>	$:$	:

* predefined in LᴬTₑX but not in TₑX.

These symbols can be negated by preceding them with \not.

Example: \not\equiv produces $\not\equiv$.

Table B.7. Relation symbols.

$\leftarrow$	\leftarrow (or \gets)	$\longleftarrow$	\longleftarrow
$\Leftarrow$	\Leftarrow	$\Longleftarrow$	\Longleftarrow
$\rightarrow$	\rightarrow (or \to)	$\longrightarrow$	\longrightarrow
$\Rightarrow$	\Rightarrow	$\Longrightarrow$	\Longrightarrow
$\leftrightarrow$	\leftrightarrow	$\longleftrightarrow$	\longleftrightarrow
$\Leftrightarrow$	\Leftrightarrow	$\Longleftrightarrow$	\Longleftrightarrow†
$\mapsto$	\mapsto	$\longmapsto$	\longmapsto
$\hookleftarrow$	\hookleftarrow	$\hookrightarrow$	\hookrightarrow
$\leftharpoonup$	\leftharpoonup	$\rightharpoonup$	\rightharpoonup
$\leftharpoondown$	\leftharpoondown	$\rightharpoondown$	\rightharpoondown
$\rightleftharpoons$	\rightleftharpoons	$\leadsto$	\leadsto*
$\uparrow$	\uparrow	$\downarrow$	\downarrow
$\Uparrow$	\Uparrow	$\Downarrow$	\Downarrow
$\updownarrow$	\updownarrow	$\Updownarrow$	\Updownarrow
$\nwarrow$	\nwarrow	$\nearrow$	\nearrow
$\swarrow$	\swarrow	$\searrow$	\searrow

* predefined in LᴬTₑX but not in TₑX.

† also \iff, which puts an extra thick space on either side.

Table B.8. Arrow symbols.

†	\dag	§	\S	©	\copyright
‡	\ddag	¶	\P	£	\pounds*
LATEX	\LaTeX*				

...	\ldots#	···	\cdots	⋮	\vdots
⋰	\ddots				
ℵ	\aleph	′	\prime	∀	\forall
ℏ	\hbar	∅	\emptyset	∃	\exists
ı	\imath	∇	\nabla	¬	\neg (or \lnot)
ȷ	\jmath	√	\surd	♭	\flat
ℓ	\ell	⊤	\top	♮	\natural
℘	\wp	⊥	\bot	♯	\sharp
ℜ	\Re	\	\backslash	∠	\angle
ℑ	\Im	∂	\partial	℧	\mho*
.	.	∞	\infty	□	\Box*
◇	\Diamond*	△	\triangle	♣	\clubsuit
◇	\diamondsuit	♡	\heartsuit	♠	\spadesuit

* predefined in LATEX but not in TEX.
# or \dots.

Table B.9. Miscellaneous symbols.

∑	\sum	∩	\bigcap	⊙	\bigodot
∏	\prod	∪	\bigcup	⊗	\bigotimes
⨿	\coprod	⊔	\bigsqcup	⊕	\bigoplus
∫	\int	⋁	\bigvee	⊎	\biguplus
∮	\oint	⋀	\bigwedge		

Table B.10. Variable-sized symbols.

\arccos	\cos	\csc	\exp	\ker	\limsup	\min	\sinh
\arcsin	\cosh	\deg	\gcd	\lg	\ln	\Pr	\sup
\arctan	\cot	\det	\hom	\lim	\log	\sec	\tan
\arg	\coth	\dim	\inf	\liminf	\max	\sin	\tanh

Table B.11. Log-like symbols.

(	(		)	)	
[	[(or \lbrack)		]	] (or \rbrack)	
{	\{		}	\}	
↑	\uparrow		↓	\downarrow	
⇑	\Uparrow		⇓	\Downarrow	
↕	\updownarrow		⇕	\Updownarrow	
⌊	\lfloor		⌋	\rfloor	
⌈	\lceil		⌉	\rceil	
⟨	\langle		⟩	\rangle	
/	/		\	\backslash	
|	| (or \vert)		‖	\\| (or \Vert)	

Table B.12. Delimiters.

⎱	\rmoustache	⎰	\lmoustache	⎱	\rgroup	⎧	\lgroup
|	\arrowvert	‖	\Arrowvert	|	\bracevert		

Table B.13. Large delimiters.

$\widetilde{abc}$	\widetilde{abc}	$\widehat{abc}$	\widehat{abc}
$\overleftarrow{abc}$	\overleftarrow{abc}	$\overrightarrow{abc}$	\overrightarrow{abc}
$\overline{abc}$	\overline{abc}	$\underline{abc}$	\underline{abc}
$\overbrace{abc}$	\overbrace{abc}	$\underbrace{abc}$	\underbrace{abc}
$\sqrt{abc}$	\sqrt{abc}	$\sqrt[n]{abc}$	\sqrt[n]{abc}*
f'	f'	$\frac{abc}{xyz}$	\frac{abc}{xyz}*

* predefined in LATEX but not in TEX.

Table B.14. Some other constructions.

[]	\!	negative thin space* (normally $-1/6$ quad)	
][	\,	thin space (normally $1/6$ of a quad)	
] [	\:†	medium space* (normally $2/9$ of a quad)	
] [	\;	thick space* (normally $5/18$ of a quad)	
] [	\quad	quad	
] [	\qquad	qquad (two quads)	

      * can appear only in math mode.

† the medium space is \: in LATEX but \> in TEX.

Table B.15. Spacing.

Appendix C
GNU Emacs—The Sixty+ Most Useful Commands

^x means hold down the control key and press x.

Esc-x means press Esc, release it, then press x (or hold down the meta key and press x).

Use tab key for filename completion in the minibuffer.

General			
^x ^c	exit	^g	panic key—aborts current situation
^x u	undo	^z	suspend Emacs (fg to restart)

Cursor Motion			
^n	next line	^p	previous line
^a	beginning of line	^e	end of line
^f	forward character	^b	back character
Esc-f	forward word	Esc-b	back word
Esc-<	beginning of buffer	Esc->	end of buffer
^v	next page	Esc-v	previous page
^l	redraw screen, centring line		

Deletion		
^d	delete next character	Esc-d delete next word

Files	
^x ^f	load file
^x i	read file into current buffer
^x ^s	save file
^x s	save all buffers
^x ^w	write named file
Esc-x write-region	save region

Buffers			
^x b	switch to another buffer	^x ^b	list all buffers
^x ^k	kill buffer		

Search and Replace	
^s	incremental search forward*
^r	incremental search backward*
Esc-%	query replace

* terminate with Esc (leave point on found item) or
^g (point remains where it was at start of search).

Cut and Paste (region = from mark to point)	
^k	kill to end of line
^k ^k	kills to end of line then next new line
^w	kill region
Esc-w	copy region to kill ring
^y	yank from kill ring
Esc-y	yank pop (use after ^y)
^@ or ^space	set mark
^x ^x	swap point and mark
^u ^@	goto previous mark
Esc-h	mark paragraph
^x h	mark buffer

Transposing, Capitalizing, Help, etc.			
^t	transpose characters	Esc-t	transpose words
Esc-l	lowercase word	Esc-u	uppercase word
^x ^u	uppercase region	^x ^l	lowercase region
^q	insert literal	^h k	describe key (help)
^h l	show last 100 chars typed	^h t	Emacs tutorial
Esc-q	reformat paragraph	^x ^t	transpose lines

Windows	
^x 2	divide screen into two windows
^x o	switch to other window
^x 1	current window becomes only window

Mail	
^x m	mail buffer
^c ^c	send mail, select other buffer
^c ^s	send mail, leave mail buffer open

Macros	
^x (	start recording keyboard macro
^x)	end recording keyboard macro
^x e	play keyboard macro

Shell	
Esc-!	shell
Esc-|	shell with region as input
Esc-x shell	start shell buffer

Appendix D
Mathematical Organizations

American Mathematical Society
P.O. Box 6248
Providence, Rhode Island 02940-6248
USA
tel: 401-455-4000
fax: 401-331-3842
email: ams@math.ams.org

Canadian Mathematical Society
577 King Edward, Suite 109
C. P. 450, Succursale A
Ottawa, Ontario
K1N 6N5, Canada
tel: 613-564-2223
fax: 613-565-1539
email: exsmc@acadvm1.uottawa.ca

Edinburgh Mathematical Society
University of Edinburgh
James Clerk Maxwell Building
The King's Buildings
Mayfield Road
Edinburgh
EH9 3JZ
Scotland
tel: 031 650 5039
fax: 031 650 6553
email: jmartin@ed.ac.uk

European Mathematical Society
Department of Mathematics
University of Helsinki
P.O. Box 4
Hallituskatu 15
SF-00014
Finland
tel: 358-0-191 2883
fax: 358-0-191 3213
email: makelainen@cc.helsinki.fi

The Institute of Mathematics and Its Applications
16 Nelson Street
Southend-on-Sea
Essex
SS1 1EF
England
tel: 0702 354020
fax: 0702 354111

London Mathematical Society
Burlington House
Piccadilly
London
W1V 0NL
England
tel: 071 437 5377
fax: 071 439 4629
email: lms@oak.cc.kcl.ac.uk

Mathematical Association
259 London Road
Leicester
LE2 3BE
England
tel: 0533 703877

Mathematical Association of America
1529 18th Street, N.W.
Washington, D.C. 20036-1375
USA
tel: 202-387-5200
fax: 202-265-2384
email: maahq@maa.org

The Society for Industrial and Applied Mathematics
3600 University City Science Center
Philadelphia, Pennsylvania 19104-2688
USA
tel: 215-382-9800
fax: 215-386-7999
email: siam@siam.org

Appendix E
Winners of Prizes for Expository Writing

This appendix is based on lists supplied by the Mathematical Association of America.

Winners of the Chauvenet Prize

Named after William Chauvenet (1820–1870), a professor of mathematics in the United States Navy, this prize is awarded for a "noteworthy paper published in English, such as will come within the range of profitable reading for a member of the Mathematical Association of America." The first twenty-four prize-winning papers (Bliss (1924)–Zalcman (1974)) are collected in the two volume *Chauvenet Papers* [1] (which, usefully, are indexed).

1925

Gilbert Ames Bliss, *Algebraic functions and their divisors*, Ann. Math., 26, 1924, pp. 95–124.

1929

T. H. Hildebrandt, *The Borel theorem and its generalizations*, Bull. Amer. Math. Soc., 32, 1926, pp. 423–474.

1932

G. H. Hardy, *An introduction to the theory of numbers*, Bull. Amer. Math. Soc., 35, 1929, pp. 778–818.

1935

Dunham Jackson, *Series of orthogonal polynomials*, Ann. Math., 2 (34), 1933, pp. 527–545; *Orthogonal trigonometric sums*, Ann. Math., 2 (34), 1933, pp. 799–814; *The convergence of Fourier series*, Amer. Math. Monthly, 1934.

1938

G. T. Whyburn, *On the structure of continua*, Bull. Amer. Math. Soc., 42, 1936, pp. 49–73.

1941

Saunders MacLane, *Modular fields*, Amer. Math. Monthly, 47 (5), 1940, pp. 259–274; *Some recent advances in algebra*, Amer. Math. Monthly, 46 (1), 1939, pp. 3–19.

1944

Robert H. Cameron, *Some introductory exercises in the manipulation of Fourier transforms*, National Mathematics Magazine, 1941, pp. 331–356.

1947

Paul R. Halmos, *The foundations of probability*, Amer. Math. Monthly, 51 (9), 1944, pp. 493–510.

1950

Mark Kac, *Random walk and the theory of Brownian motion*, Amer. Math. Monthly, 54 (7), 1947, pp. 369–391.

1953

E. J. McShane, *Partial orderings and Moore–Smith limits*, Amer. Math. Monthly, 59 (1), 1952, pp. 1–11.

1956

R. H. Bruck, *Recent advances in the foundations of Euclidean plane geometry*, Amer. Math. Monthly, 52 (7, Part II), 1955, pp. 2–17.

1960

Cornelius Lanczos, *Linear systems in self-adjoint form*, Amer. Math. Monthly, 65 (9), 1958, pp. 665–679.

1963

Philip J. Davies, *Leonhard Euler's integral: A historical profile of the gamma function*, Amer. Math. Monthly, 66 (10), 1959, pp. 849–869.

1964

Leon A. Henkin, *Are logic and mathematics identical?*, Science, 138 (3542), 1962, pp. 788–794.

1965

Jack K. Hale and Joseph P. LaSalle, *Differential equations: Linearity vs. nonlinearity*, SIAM Rev., 5 (3), 1963, pp. 249–272.

1966

No award

1967

Guido L. Weiss, *Harmonic analysis*, MAA Stud. Math., 3, 1965, pp. 124–178.

1968

Mark Kac, *Can one hear the shape of a drum?*, Amer. Math. Monthly, 73 (4, Part II), Slaught Papers No. 11, 1966, pp. 1–23.

1969

No award

1970

Shiing-Shen Chern, *Curves and surfaces in Euclidean space*, MAA Stud. in Math., 1967, pp. 16–56.

1971

Norman Levinson, *A motivated account of an elementary proof of the prime number theorem*, Amer. Math. Monthly, 76 (2), 1969, pp. 225–245.

1972

Jean François Trèves, *On local solvability of linear partial differential equations*, Bull. Amer. Math. Soc., 76, 1970, pp. 552–571.

1973

Carl D. Olds, *The simple continued fraction expansion of e*, Amer. Math. Monthly, 77 (9), 1970, pp. 968–974.

1974

Peter D. Lax, *The formation and decay of shock waves*, Amer. Math. Monthly, 79 (3), 1972, pp. 227–241.

1975

Martin D. Davis and Reuben Hersh, *Hilbert's 10th problem*, Scientific American, 229 (5), November 1973, pp. 84–91.

1976

Lawrence Zalcman, *Real proofs of complex theorems (and vice versa)*, Amer. Math. Monthly, 81 (2), 1974, pp. 115–137.

1977

W. Gilbert Strang, *Piecewise polynomials and the finite element method*, Bull. Amer. Math. Soc., 79 (6), 1973, 1128–1137.

1978

Shreeram S. Abhyankar, *Historical ramblings in algebraic geometry and related algebra*, Amer. Math. Monthly, 83 (6), 1976, pp. 409–448.

1979

Neil J. A. Sloane, *Error-correcting codes and invariant theory: New applications of a nineteenth-century technique*, Amer. Math. Monthly, 84 (2), 1977, pp. 82–107.

1980

Heinz Bauer, *Approximation and abstract boundaries*, Amer. Math. Monthly, 85 (8), 1978, pp. 632–647.

1981

Kenneth I. Gross, *On the evolution of noncommutative harmonic analysis*, Amer. Math. Monthly, 85 (7), 1978, pp. 525–548.

1982

No award

1983

No award

1984

R. Arthur Knoebel, *Exponentials reiterated*, Amer. Math. Monthly, 88 (4), 1981, pp. 235–252.

1985

Carl Pomerance, *Recent developments in primality testing*, Mathematical Intelligencer, 3 (3), 1981, pp. 97–105.

1986

George Miel, *Of calculations past and present: The Archimedean algorithm*, Amer. Math. Monthly, 90 (17), 1983, pp. 17–35.

1987

James H. Wilkinson, *The perfidious polynomial*, in Gene H. Golub, ed., Studies in Numerical Analysis, vol. 24 of Studies in Mathematics, The Mathematical Association of America, Washington, D.C., 1984, pp. 1–28.

1988

Steve Smale, *On the efficiency of algorithms of analysis*, Bull. Amer. Math. Soc., 13 (2), 1985, pp. 87–121.

1989

Jacob Korevaar, *Ludwig Bieberbach's conjecture and its proof by Louis de Branges*, Amer. Math. Monthly, 93 (7), 1986, pp. 505–514.

1990

W. B. R. Lickorish and Kenneth C. Millett, *The new polynomial invariants of knots and links*, Math. Mag., 61 (1), 1988, pp. 3–23.

1991

Steven G. Krantz, *What is several complex variables?*, Amer. Math. Monthly, 94 (3), 1987, pp. 236–256.

1992

J. M. Borwein, P. B. Borwein, and D. H. Bailey, *Ramanujan, modular equations, and approximations to Pi or how to compute one billion digits of Pi*, Amer. Math. Monthly, 96 (3), 1989, pp. 201–219.

Winners of the Lester R. Ford Award

An award for articles in the American Mathematical Monthly.

1965

R. H. Bing, *Spheres in E^3*, Amer. Math. Monthly, 71 (4), 1964, pp. 353–364.

Louis Brand, *A division algebra for sequences and its associated operational calculus*, Amer. Math. Monthly, 71 (7), 1964, pp. 719–728.

R. G. Kuller, *Coin tossing, probability, and the Weierstrass approximation theorem*, Math. Mag., 37, 1964, pp. 262–265.

R. D. Luce, *The mathematics used in mathematical psychology*, Amer. Math. Monthly, 71 (4), 1964, pp. 364–378.

Hartley Rogers, Jr., *Information theory*, Math. Mag., 37, 1964, pp. 63–78.

Elmer Tolsted, *An elementary derivation of the Cauchy, Hölder, and Minkowski inequalities from Young's inequality*, Math. Mag., 37, 1964, pp. 2–12.

1966

C. B. Allendoerfer, *Generalizations of theorems about triangles*, Math. Mag., 38, 1965, pp. 253–259.

Peter D. Lax, *Numerical solution of partial differential equations*, Amer. Math. Monthly, 72 (2, Part II), Slaught Papers No. 10, 1965, pp. 74–84.

Marvin Marcus and Henryk Minc, *Permanents*, Amer. Math. Monthly, 72 (6), 1965, pp. 577–591.

1967

Wai-Kai Chen, *Boolean matrices and switching nets*, Math. Mag., 36, 1966, pp. 1–8.

D. R. Fulkerson, *Flow networks and combinatorial operations research*, Amer. Math. Monthly, 73 (2), 1966, pp. 115–138.

Mark Kac, *Can one hear the shape of a drum?*, Amer. Math. Monthly, 73 (4, Part II), Slaught Papers No. 11, 1966, pp. 1–23.

M. Z. Nashed, *Some remarks on variations and differentials*, Amer. Math. Monthly, 73 (4, Part II), Slaught Papers No. 11, 1966, pp. 63–76.

P. B. Yale, *Automorphisms of the complex numbers*, Math. Mag., 39, 1966, pp. 135–141.

1968

Frederic Cunningham, Jr., *Taking limits under the integral sign*, Math. Mag., 40, 1967, pp. 179–186.

W. F. Newns, *Functional dependence*, Amer. Math. Monthly, 74 (8), 1967, pp. 911–920.

Daniel Pedoe, *On a theorem in geometry*, Amer. Math. Monthly, 74 (6), 1967, pp. 627–640.

Keith L. Phillips, *The maximal theorems of Hardy and Littlewood*, Amer. Math. Monthly, 74 (6), 1967, pp. 648–660.

F. V. Waugh and Margaret W. Maxfield, *Side-and-diagonal numbers*, Math. Mag., 40, 1967, pp. 74–83.

Hans J. Zassenhaus, *On the fundamental theorem of algebra*, Amer. Math. Monthly, 74 (5), 1967, pp. 485–497.

1969

Harley Flanders, *A proof of Minkowski's inequality for convex curves*, Amer. Math. Monthly, 75 (6), 1968, pp. 581–593.

George E. Forsythe, *What to do till the computer scientist comes*, Amer. Math. Monthly, 75 (5), 1968, pp. 454–462.

M. F. Neuts, *Are many 1-1-functions on the positive integers onto?*, Math. Mag., 41, 1968, pp. 103–109.

Pierre Samuel, *Unique factorization*, Amer. Math. Monthly, 75 (9), 1968, pp. 945–952.

Hassler Whitney, *The mathematics of physical quantities*, I *and* II, Amer. Math. Monthly, 75 (2,3), 1968, pp. 115–138 and 227–256.

Albert Wilansky, *Spectral decomposition of matrices for high school students*, Math. Mag., 41, 1968, pp. 51–59.

1970

Henry L. Alder, *Partition identities—from Euler to the present*, Amer. Math. Monthly, 76 (7), 1969, pp. 733–746.

Ralph P. Boas, *Inequalities for the derivatives of polynomials*, Math. Mag., 42, 1969, pp. 165–174.

W. A. Coppel, *J. B. Fourier—on the occasion of his two hundredth birthday*, Amer. Math. Monthly, 76 (5), 1969, pp. 468–483.

Norman Levinson, *A motivated account of an elementary proof of the prime number theorem*, Amer. Math. Monthly, 3, 1969, pp. 225–245.

John Milnor, *A problem in cartography*, Amer. Math. Monthly, 10, 1969, pp. 1101–1112.

Ivan Niven, *Formal power series*, Amer. Math. Monthly, 8, 1969, pp. 871–889.

1971

Jean A. Dieudonné, *The work of Nicholas Bourbaki*, Amer. Math. Monthly, 77 (2), 1970, pp. 134–145.

George E. Forsythe, *Pitfalls in computation, or why a math book isn't enough*, Amer. Math. Monthly, 77 (9), 1970, pp. 931–956.

Paul R. Halmos, *Finite-dimensional Hilbert spaces*, Amer. Math. Monthly, 77 (5), 1970, pp. 457–464.

Eric Langford, *A problem in geometric probability*, Math. Mag., 43, 1970, pp. 237–244.

P. V. O'Neil, *Ulam's conjecture and graph reconstructions*, Amer. Math. Monthly, 77 (1), 1970, pp. 35–43.

Olga Taussky, *Sums of squares*, Amer. Math. Monthly, 77 (8), 1970, pp. 805–830.

1972

G. D. Chakerian and L. H. Lange, *Geometric extremum problems*, Math, Mag., 44, 1971, pp. 57–69.

P. M. Cohn, *Rings of fractions*, Amer. Math. Monthly, 78 (6), 1971, pp. 596–615.

Frederic Cunningham, Jr., *The Kakeya problem for simply connected and for star-shaped sets*, Amer. Math. Monthly, 78 (2), 1971, pp. 114–129.

W. J. Ellison, *Waring's problem*, Amer. Math. Monthly, 78 (1), 1971, pp. 10–36.

Leon Henkin, *Mathematical foundations for mathematics*, Amer. Math. Monthly, 78 (5), 1971, pp. 463–487.

Victor Klee, *What is a convex set?*, Amer. Math. Monthly, 78 (6), 1971, pp. 616–631.

1973

Jean A. Dieudonné, *The historical development of algebraic geometry*, Amer. Math. Monthly, 79 (8), 1972, pp. 827–866.

Samuel Karlin, *Some mathematical models of population genetics*, Amer. Math. Monthly, 79 (7), 1972, pp. 699–739.

Peter D. Lax, *The formation and decay of shock waves*, Amer. Math. Monthly, 79 (3), 1972, pp. 227–241.

Thomas L. Saaty, *Thirteen colorful variations on Guthrie's four-color conjecture*, Amer. Math. Monthly, 79 (1), 1972, pp. 2–43.

Lynn A. Steen, *Conjectures and counterexamples in metrization theory*, Amer. Math. Monthly, 79 (2), 1972, pp. 113–132.

R. L. Wilder, *History in the mathematics curriculum: Its status, quality, and function*, Amer. Math. Monthly, 79 (5), 1972, pp. 479–495

1974

Patrick Billinglsey, *Prime numbers and Brownian motion*, Amer. Math. Monthly, 80 (10), 1973, pp. 1099–1115.

Garrett Birkhoff, *Current trends in algebra*, Amer. Math. Monthly, 80 (7), 1973, pp. 760–782.

Martin Davis, *Hilbert's tenth problem is unsolvable*, Amer. Math. Monthly, 80 (3), 1973, pp. 233–269.

I. J. Schoenberg, *The elementary cases of Landau's problem of inequalities between derivatives*, Amer. Math. Monthly, 80 (2), 1973, pp. 121–158.

Lynn A. Steen, *Highlights in the history of spectral theory*, Amer. Math. Monthly, 80 (4), 1973, pp. 359–381

Robin J. Wilson, *An introduction to matroid theory*, Amer. Math. Monthly, 80 (5), 1973, pp. 500–525.

1975

Raymond Ayoub, *Euler and the zeta function*, Amer. Math. Monthly, 81 (10), 1974, pp. 1067–1086.

James Callahan, *Singularities and plane maps*, Amer. Math. Monthly, 81 (3), 1974, pp. 211–240.

Donald E. Knuth, *Computer science and its relation to mathematics*, Amer. Math. Monthly, 81 (4), 1974, pp. 323–343.

Johannes C. C. Nitsche, *Plateau's problems and their modern ramifications*, Amer. Math. Monthly, 81 (9), 1974, pp. 945–968.

S. K. Stein, *Algebraic tiling*, Amer. Math. Monthly, 81 (5), 1974, pp. 445–462.

Lawrence Zalcman, *Real proofs of complex theorems (and vice versa)*, Amer. Math. Monthly, 81 (2), 1974, pp. 115–137.

1976

M. L. Balinski and H. P. Young, *The quota method of apportionment*, Amer. Math. Monthly, 82 (7), 1975, pp. 701–730.

E. A. Bender and J. R. Goldman, *On the applications of Mobius inversion in combinatorial analysis*, Amer. Math. Monthly, 82 (8), 1975, pp. 789–803.

Branko Grünbaum, *Venn diagrams and independent families of sets*, Math. Mag., 48, 1975, pp. 12–23.

J. E. Humphreys, *Representations of SL(2,p)*, Amer. Math. Monthly, 82 (1), 1975, pp. 21–39.

J. B. Keller and D. W. McLaughlin, *The Feynman integral*, Amer. Math. Monthly, 82 (5), 1975, pp. 451–465.

J. J. Price, *Topics in orthogonal functions*, Amer. Math. Monthly, 82 (6), 1975, pp. 594–609.

1977

Shreeram S. Abhyankar, *Historical ramblings in algebraic geometry and related algebra*, Amer. Math. Monthly, 83 (6), 1976, pp. 409–448.

Joseph B. Keller, *Inverse problems*, Amer. Math. Monthly, 83 (2), 1976, pp. 107–118.

D. S. Passman, *What is a group ring?*, Amer. Math. Monthly, 83 (3), 1976, pp. 173–185.

James P. Jones, Diahachiro Sato, Hideo Wada, and Douglas Wiens, *Diophantine representation of the set of prime numbers*, Amer. Math. Monthly, 83 (6), 1976, pp. 449–464.

J. H. Ewing, W. H. Gustafson, P. R. Halmos, S. H. Moolgavkar, W. H. Wheeler, and W. P. Ziemer, *American mathematics from 1940 to the day before yesterday*, Amer. Math. Monthly, 83 (7), 1976, pp. 503–516.

1978

Ralph P. Boas, Jr., *Partial sums of infinite series, and how they grow*, Amer. Math. Monthly, 84 (4), 1977, pp. 237–258.

Louis H. Kauffman and Thomas F. Banchoff, *Immersions and mod-2 quadratic forms*, Amer. Math. Monthly, 84 (3), 1977, pp. 168–185.

Neil J. A. Sloane, *Error-correcting codes and invariant theory: New applications of a nineteenth-century technique*, Amer. Math. Monthly, 84 (2), 1977, pp. 82–107.

1979

Bradley Efron, *Controversies in the foundations of statistics*, Amer. Math. Monthly, 85 (4), 1978, pp. 231–246.

Ned Glick, *Breaking records and breaking boards*, Amer. Math. Monthly, 85 (1), 1978, pp. 2–26.

Kenneth I. Gross, *On the evolution of noncommutative harmonic analysis*, Amer. Math. Monthly, 85 (7), 1978, pp. 525–548.

Lawrence A. Shepp and Joseph B. Kruskal, *Computerized tomography: The new medical X-ray technology*, Amer. Math. Monthly, 85 (6), 1978, pp. 420–439.

1980

Desmond P. Fearnley-Sander, *Hermann Grassmann and the creation of linear algebra*, Amer. Math. Monthly, 86 (10), 1979, pp. 809–817.

David Gale, *The game of hex and the Brouwer fixed-point theorem*, Amer. Math. Monthly, 86 (10), 1979, pp. 818–826.

Karel Hrbacek, *Nonstandard set theory*, Amer. Math. Monthly, 86 (8), 1979, pp. 659–677.

Cathleen S. Morawetz, *Nonlinear conservation equations*, Amer. Math. Monthly, 86 (4), 1979, pp. 284–287.

Robert Osserman, *Bonnesen-style isoperimetric inequalities*, Amer. Math. Monthly, 86 (1), 1979, pp. 1–29.

1981

R. Creighton Buck, *Sherlock Holmes in Babylon*, Amer. Math. Monthly, 87 (5), 1980, pp. 335–345.

Brune H. Pourciau, *Modern multiplier rules*, Amer. Math. Monthly, 87 (6), 1980, pp. 433-452.

Alan H. Schoenfeld, *Teaching problem-solving skills*, Amer. Math. Monthly, 87 (10), 1980, pp. 794–805.

Edward R. Swart, *The philosophical implications of the four-color problem*, Amer. Math. Monthly, 87 (9), 1980, pp. 697–707.

Lawrence A. Zalcman, *Offbeat integral geometry*, Amer. Math. Monthly, 87 (3), 1980, pp. 161–175.

1982

Philip J. Davis, *Are there coincidences in mathematics?*, Amer. Math. Monthly, 88 (5), 1981, pp. 311–320.

R. Arthur Knoebel, *Exponentials reiterated*, Amer. Math. Monthly, 88 (4), 1981, pp. 235–252.

1983

Robert F. Brown, *The fixed point property and Cartesian products*, Amer. Math. Monthly, 89 (9), 1982, pp. 654–678.

Tony Rothman, *Genius and biographers: The fictionalization of Evariste Galois*, Amer. Math. Monthly, 89 (2), 1982, pp. 84–106.

Robert S. Strichartz, *Radon inversion—variations on a theme*, Amer. Math. Monthly, 89 (6), 1982, pp. 377–384 and 420–423 (solutions of problems).

1984

Judith V. Grabiner, *Who gave you the epsilon? Cauchy and the origins of rigorous calculus*, Amer. Math. Monthly, 90 (3), 1983, pp. 185–194.

Roger Howe, *Very basic Lie theory*, Amer. Math. Monthly, 90 (9), 1983, pp. 600–623.

John Milnor, *On the geometry of the Kepler problem*, Amer. Math. Monthly, 90 (6), 1983, pp. 353–365.

Joel Spencer, *Large numbers and unprovable theorems*, Amer. Math Monthly, 90 (10), 1983, pp. 669–675.

William C. Waterhouse, *Do symmetric problems have symmetric solutions?*, Amer. Math. Monthly, 90 (6), 1983, pp. 378–387.

1985

John D. Dixon, *Factorization and primality tests*, Amer. Math. Monthly, 91 (6), 1984, pp. 333–352.

Donald G. Saari and John B. Urenko, *Newton's method, circle maps, and chaotic motion*, Amer. Math. Monthly, 91 (1), 1984, pp. 3–17.

1986

Jeffrey C. Lagarias, *The $3x+1$ problem and its generalizations*, Amer. Math. Monthly, 92 (1), 1985, pp. 3–23.

Michael E. Taylor, *Review of Lars Hörmander's "The Analysis of linear partial differential operations, I and II"*, Amer. Math. Monthly, 92 (10), 1985, pp. 745–749.

1987

Stuart S. Antman, *Review of Ann Hibler Koblitz's "A convergence of lives— Sophia Kovalevskaia: Scientist, Writer, Revolutionary"*, Amer. Math. Monthly, 93 (2), 1986, pp. 139–144.

Joan Cleary, Sidney A. Morris and David Yost, *Numerical geometry— numbers for shapes*, Amer. Math. Monthly, 93 (4), 1986, pp. 260–275.

Howard Hiller, *Crystallography and cohomology of groups*, Amer. Math. Monthly, 93 (10), 1986, pp. 765–779.

Jacob Korevaar, *Ludwig Bieberbach's conjecture and its proof by Louis de Branges*, Amer. Math. Monthly, 93 (7), 1986, pp. 505–514.

Peter M. Neumann, *Review of Harold M. Edwards' "Galois Theory"*, Amer. Math. Monthly, 93 (5), 1986, pp. 407–411.

1988

James F. Epperson, *On the Runge example*, Amer. Math. Monthly, 94 (4), 1987, pp. 329–341.

Stan Wagon, *Fourteen proofs of a result about tiling a rectangle*, Amer. Math. Monthly, 94 (7), 1987, pp. 601–617.

1989

Richard K. Guy, *The strong law of small numbers*, Amer. Math. Monthly, 95 (8), 198, pp. 697–712.

Gert Almkvist and Bruce Berndt, *Gauss, Landen, Ramanujan, the arithmetic-geometric mean, ellipses, π and the "Ladies Diary"*, Amer. Math. Monthly, 95 (7), 1988, pp. 585–608.

1990

Jacob E. Goodman, János Pach and Chee K. Yap, *Mountain climbing, ladder moving and the ring-width of a polygon*, Amer. Math. Monthly, 96 (6), 1989, pp. 494–510.

Doron Zeilberger, *Kathy O'Hara's constructive proof of the unimodality of the Gaussian polynomials*, Amer. Math. Monthly, 96 (7), 1989, pp. 590–602.

1991

Marcel Berger, *Convexity*, Amer. Math. Monthly, 97 (8), 1990, pp. 650–678.

Ronald L. Graham and Frances Yao, *A whirlwind tour of computational geometry*, Amer. Math. Monthly, 97 (8), 1990, pp. 687–701.

Joyce Justicz, Edward R. Scheinerman and Peter M. Winkler, *Random intervals*, Amer. Math. Monthly, 97 (10), 1990, pp. 881–889.

Winners of the George Polya Award

An award for articles in the College Mathematics Journal—formerly the Two-Year College Mathematics Journal.

1977

Julian Weisglass, *Small groups: An alternative to the lecture method*, 7 (1), 1976, pp. 15–20.

Anneli Lax, *Linear algebra, A potent tool*, 7 (2), 1976, pp. 3–15.

1978

Allen H. Holmes, Walter J. Sanders and John W. LeDuc, *Statistical inference for the general education student—it can be done*, 8 (4), 1977, pp. 223–230.

Freida Zames, *Surface area and the cylinder area paradox*, 8 (4), 1977, pp. 207–211.

1979

Richard L. Francis, *A note on angle construction*, 9 (2), 1978, pp. 75–80.

Richard Plagge, *Fractions without quotients: Arithmetic of repeating decimals*, 9 (1), 1978, pp. 11–15.

1980

Hugh F. Ouellette and Gordon Bennett, *The discovery of a generalization*, 10 (2), 1979, pp. 100–106.

Robert Nelson, *Pictures, probability and paradox*, 10 (3), 1979, pp. 182–190.

1981

Gulbank D. Chakerian, *Circles and spheres*, 11 (1), 1980, pp. 26–41.

Dennis D. McCune, Robert G. Dean and William D. Clark, *Calculators to motivate infinite composition of functions*, 11 (3), 1980, pp. 189–195.

1982

John A. Mitchem, *On the history and solution of the four-color problem*, 12 (2), 1981, pp. 108–116.

Peter L. Renz, *Mathematical proof: What it is and what it ought to be*, 12 (2), 1981, pp. 83–103.

1983

Douglas R. Hofstadter, *Analogies and metaphors to explain Godel's theorem*, 13 (2), 1982, pp. 98–114.

Paul R. Halmos, *The thrills of abstraction*, 13 (4), September 1982, pp. 243–251.

Warren Page and V. N. Murty, *Nearness relations among measures of central tendency and dispersion, Part 1*, 13 (5), 1982, pp. 315–327.

1984

Ruma Falk and Maya Bar-Hillel, *Probabilistic dependence between events*, 14 (3), 1983, pp. 240–247.

Richard J. Trudeau, *How big is a point?*, 14 (4), 1983, pp. 295–300.

1985

Anthony Barcellos, *The fractal geometry of Mandelbrot*, 15 (2), 1984, pp. 98–114.

Kay W. Dundas, *To build a better box*, 15 (1), 1984, pp. 30–36.

1986

Philip J. Davis, *What do I know? A study of mathematical self-awareness*, 16 (1), 1985, pp. 22–41.

1987

Constance Reid, *The autobiography of Julia Robinson*, 17 (1), 1986, pp. 3–21.

Irl C. Bivens, *What a tangent line is when it isn't a limit*, 17 (2), 1986, pp. 133–143.

1988

Dennis M. Luciano and Gordon M. Prichett, *From Caesar ciphers to public-key cryptosystems*, 18 (1), 1987, pp. 2–17.

V. Frederick Rickey, *Isaac Newton: Man, myth and mathematics*, 18 (5), 1987, pp. 362–389.

1989

Edward Rozema, *Why should we pivot in Gaussian elimination?*, 19 (1), 1988, pp. 63–72.

Beverly L. Brechner and John C. Mayer, *Antoine's necklace or how to keep a necklace from falling apart*, 19 (4), 1988, pp. 306–320.

1990

Israel Kleiner, *Evolution of the function concept: A brief survey*, 20 (4), 1989, pp. 282–300.

Richard D. Neidinger, *Automatic differentiation and APL*, 20 (3), 1989, pp. 238–251.

1991

William B. Gearhart and Harris S. Shultz, *The function* $\sin x/x$, 21 (2), 1990, pp. 90–99.

Mark Schilling, *The longest run of heads*, 21 (3), 1990, pp. 196–207.

Winners of the Carl B. Allendoerfer Award

An award for articles in Mathematics Magazine.

1977

Joseph A. Gallian, *The search for finite groups*, 49 (4), 1976, pp. 163–179.

B. L. van der Waerden, *Hamilton's discovery of quaternions*, 49 (5), 1976, pp. 227–234.

1978

Geoffrey C. Shephard and Branko Grünbaum, *Tilings by regular polygons*, 50 (5), 1977, pp. 227–247.

David A. Smith, *Human population growth: Stability or explosion*, 50 (4), 1977, pp. 186–197.

1979

Doris W. Scattschneider, *Tiling the plane with congruent pentagons*, 51 (1), 1978, pp. 29–44.

Bruce C. Berndt, *Ramanujan's Notebooks*, 51 (3), 1978, pp. 147–164.

1980

Ernst Snapper, *The three crises in mathematics: Logicism, intuitionism, and formalism*, 52 (4), 1979, pp. 207–216.

Victor L. Klee, Jr., *Some unsolved problems in plane geometry*, 52 (3), 1979, pp. 131–145.

1981

Stephen B. Maurer, *The king chicken theorems*, 53 (2), 1980, pp. 67–80.

Donald E. Sanderson, *Advanced plane topology from an elementary standpoint*, 53 (2), 1980, pp. 81–89.

1982

J. Ian Richards, *Continued fractions without tears*, 54 (4), 1981, pp. 163–171.

Marjorie Senechal, *Which tetrahedra fill space?*, 54 (5), 1981, pp. 227–243.

1983

Donald O. Koehler, *Mathematics and literature*, 55 (2), 1982, pp. 81–95.

Clifford H. Wagner, *A generic approach to iterative methods*, 55 (5), 1982, pp. 259–273.

1984

Judith Grabiner, *The changing concept of change: The derivative from Fermat to Weierstrass*, 56 (4), 1983, pp. 195–206.

1985

Philip D. Straffin, Jr. and Bernard Grofman, *Parliamentary coalitions: A tour of models*, 57 (5), 1984, pp. 259–274.

Frederick S. Gass, *Constructive ordinal notation systems*, 57 (3), 1984, pp. 131–141.

1986

Bart Braden, *The design of an oscillating sprinkler*, 58 (1), 1985, pp. 29–38.

Saul Stahl, *The other map coloring theorem*, 58 (3), 1985, pp. 131–145.

1987

Israel Kleiner, *The evolution of group theory*, 59 (4), 1986, pp. 195–215.

Paul Zorn, *The Bieberbach conjecture*, 59 (3), 1986, pp. 131–148.

1988

Steven Galovich, *Products of sines and cosines*, 60 (2), 1987, pp. 105–113.

Bart Braden, *Polya's geometric picture of complex contour integrals*, 60 (5), 1987, pp. 321–327.

1989

Judith V. Grabiner, *The centrality of mathematics in the history of western thought*, 61 (4), 1988, pp. 220–230.

W. B. Raymond Lickorish and Kenneth C. Millett, *The new polynomial invariants of knots and links*, 61 (1), 1988, pp. 3–23.

1990

Fan K. Chung, Martin Gardner, and Ronald L. Graham, *Steiner trees on a checkerboard*, 62 (2), 1989, pp. 83–96.

Thomas Archibald, *Connectivity and smoke-rings: Green's second identity in its first fifty years*, 62 (4), 1989, pp. 219–237.

1991

Ranjan Roy, *The discovery of the series formula for π by Leibniz, Gregory and Nilakantha*, 63 (5), 1990, pp. 291–306.

1992

Israel Kleiner, *Rigor and proof in mathematics: A historical perspective*, vol. 64 (5), 1991, pp. 291–314.

G. D. Chakerian and David Logothetti, *Cube slices, pictorial triangles, and probability*, vol. 64 (4), 1991, pp. 219–241.

Winners of the MAA Book Prize, renamed Beckenbach Book Prize (1986)

1984

Charles Robert Hadlock, *Field Theory and Its Classical Problems*, Carus Monograph No. 19, 1978.

1986

Edward Packel, *The Mathematics of Games and Gambling*, MAA New Mathematical Library Series, 1981.

1989

Thomas M. Thompson, *From Error-Correcting Codes through Sphere Packings to Simple Groups*, Carus Monograph No. 21, 1984.

Winners of the Merten M. Hasse Prize (established 1986)

1987

Anthony Barcellos, *The fractal geometry of Mandelbrot*, The College Mathematics Journal, 15, 1984.

1989

Irl C. Bivens, *What a tangent is when it isn't a limit*, The College Mathematics Journal, 17, 1986.

1991

Barry A. Cipra, *An introduction to the Ising model*, Amer. Math. Monthly, 94 (10), 1987, pp. 937–959.

Glossary

abstract. A brief, self-contained summary of the contents of a paper that appears by itself at the beginning of the paper. Also a brief (written) summary of the contents of a talk.

ACM. The Association for Computing Machinery.

acknowledgements. A section preceding the references (or a footnote, or a final paragraph) in which the author thanks people or organizations for help, advice or financial support for the work described.

AMS. The American Mathematical Society.

A$_M$S-T$_E$X. A macro package for T$_E$X that makes it easier to typeset mathematical papers with T$_E$X. It gives new structures for displaying mathematical equations and comes with a special font of mathematical symbols.

anonymous ftp. A form of ftp in which the user logs on as user `anonymous` and need not type a password (though, by convention, the user's electronic mail address is typed as the password).

archive. In Unix, a single file that contains a set of other files (e.g., as manipulated with the `tar` command). Or a collection of software that is located at a particular Internet address and can be accessed by anonymous ftp.

ASCII. American Standard Code for Information Interchange. A coding system in which letters, digits, punctuation symbols and control characters are represented in seven bits by a number from 0 to 127. An eighth bit is often added to allow extra characters.

bibliography. A list of publications on a particular topic, or the reference list of a book.

B$_IB$T$_E$X. A program that cooperates with L^{A}T$_E$X in the preparation of reference lists. It makes use of `bib` files, which are databases of references in B$_IB$T$_E$X format.

citation. A reference in the text to a publication or other source, usually one that is listed in the references.

compositor. The person who typesets the text (especially in traditional printing).

Computing Reviews Classification System. A classification system for computer science. An example of an entry is G.1.3 [Numerical Analysis]: Numerical Linear Algebra—sparse and very large systems.

conjecture. A statement that the author believes to be true but for which a proof or disproof has not been found.

copy editor. A person who prepares a manuscript for typesetting by checking and correcting grammar, punctuation, spelling, style, consistency and other details.

corollary. A direct or easy consequence of a lemma, theorem or proposition.

Current Contents. A publication from the Institute for Scientific Information that provides a weekly list of journal contents pages. The Physical Sciences edition is the one in which mathematics and computer science journals appear.

DOS. Disk Operating System. Usually refers to MS-DOS (Microsoft Disk Operating System) or PC-DOS (IBM), which are computer operating systems for personal computers (PCs).

Festschrift (or festschrift). (German) A collection of writings published in honour of a scholar.

folio. A printed page number. Also a sheet of a manuscript.

ftp. File transfer protocol. A protocol for file transfer between different computers on the Internet network. Also a program for transferring files using this protocol.

Harvard system. A system of citation by author name and year, e.g., "see Knuth (1986)".

hypothesis. A statement taken as a basis for further reasoning.

IMA. The Institute of Mathematics and Its Applications, Southend-on-Sea, England. Also the Institute for Mathematics and Its Applications, University of Minnesota, Minneapolis, USA.

Internet. The worldwide network of interconnected computer networks. It provides electronic mail, file transfer, news, remote login, and other services.

ISBN. International Standard Book Number.

LaTeX. A macro package for TeX that simplifies the production of papers, books and letters, and emphasizes the logical structure of a document. It permits automatic cross-referencing and has commands for drawing pictures.

lemma. An auxiliary result needed in the proof of a theorem or proposition. May also be an independent result that does not merit the title theorem.

LMS. The London Mathematical Society.

MAA. The Mathematical Association of America.

macro. In computing, a shorthand notation for specifying a sequence of operations. For example, in typesetting this book in LaTeX I used the definition `\def\mw{mathematical writing}` and typed `\mw` whenever I wanted the phrase "mathematical writing" to appear.

manuscript. Literally, a handwritten document. More generally, any unpublished document, particularly one submitted for publication.

managing editor. A person who is in charge of the editorial activities of a publication and who supervises a staff of editors.

Mathematical Reviews. A monthly review publication run by the American Mathematical Society (AMS) and first published in 1940. Each listed paper is accompanied by a review or a reprint of the paper's abstract.

Mathematics Subject Classifications. A classification scheme published in *Mathematical Reviews* that divides mathematics into 61 sections numbered between 0 and 94, further divided into many subsections. A typical entry is 65F05 (direct methods for solving linear systems).

netlib. An electronic repository of public domain mathematical software for the scientific computing community, the main copies of which are located at Internet addresses `netlib@ornl.gov` and `netlib@research.att.com`.

offprint. See reprint.

page charges. Charges levied by a publisher to offset the cost of publishing an article. In mathematics journals payment is usually optional.

peer review. Refereeing done by peers of the author (people working in the same area). Should perhaps be called "peer refereeing", but "peer review" is standard.

proceedings. A collection of papers describing the work presented at a conference or workshop. Also may be a title for a journal: for example, *The Proceedings of the American Mathematical Society.*

proofreading. The process of checking proofs for errors (usually by comparing them with an original) and marking the errors with standard proofreading symbols.

proofs. Typeset material ready for checking and correction. *Galley proofs* are traditionally sheets that have not been broken up into pages; *page proofs* have been paginated.

proposition. Same meaning as a theorem (but possibly regarded as a lesser result).

PostScript. A page description language developed by Adobe Systems, Inc.

referee. A person who advises an editor on the suitability of a manuscript for publication.

references. The list of publications cited in the text, or those publications themselves.

reprint. A separate printing of an article that appeared in a book or journal. Often, a limited number are supplied free of charge to the author.

repro. Also known as "mechanicals". Camera-ready proofs of material to be printed.

reviewer. A person who reviews previously published or completed work. Sometimes used incorrectly as a synonym for referee.

running head. An abbreviated title that appears in the headline of pages in a published paper.

Science Citation Index. A publication from the Institute for Scientific Information that records all papers that reference an earlier paper, across all science subjects. It covers the period from 1945 to the present.

SIAM. The Society for Industrial and Applied Mathematics.

technical report. A document published by an organization for external circulation, usually as part of a series.

TEX. A system for computer typesetting of mathematics, developed by Donald Knuth at Stanford University. Also used as a verb: "to TEX a paper" is to typeset the paper in TEX.

theorem. A major result of independent interest.

thesaurus. A list of words in which each word is followed by a list of words of similar meaning or sense. The main list may be arranged by meaning (Roget's thesaurus) or alphabetically (most other thesauri).

title. "The fewest possible words that adequately describe the contents of a paper, book, poster, etc." [51]

Unix. A computer operating system developed at Bell Laboratories. Widely used on workstations and supercomputers.

widow. A short last line of a paragraph appearing at the top of a page.

Bibliography

[1] J. C. ABBOTT, ED., *The Chauvenet Papers: A Collection of Prize-Winning Expository Papers in Mathematics, Volumes 1 and 2*, Mathematical Association of America, Washington, D.C., 1978.

[2] MILTON ABRAMOWITZ AND IRENE A. STEGUN, EDS., *Handbook of Mathematical Functions with Formulas, Graphs and Mathematical Tables*, vol. 55 of Applied Mathematics Series, National Bureau of Standards, Washington, D.C., 1964. Reprinted by Dover, New York.

[3] FORMAN S. ACTON, *Numerical Methods That Work*, Harper and Row, New York, 1970. Reprinted by Mathematical Association of America, Washington, D.C., with new preface and additional problems, 1990. ISBN 0-88385-450-3.

[4] ALFRED V. AHO, BRIAN W. KERNIGHAN, AND PETER J. WEINBERGER, *The AWK Programming Language*, Addison-Wesley, Reading, Massachusetts, 1988. ISBN 0 201 07981 X.

[5] D. J. ALBERS AND G. L. ALEXANDERSON, EDS., *Mathematical People: Profiles and Interviews*, Birkhäuser, Boston, 1985.

[6] AMERICAN MATHEMATICAL SOCIETY, $A_{\mathcal{M}}S$-$\mathbb{A}T_EX$ *Version 1.1 User's Guide*, Providence, Rhode Island, 1991.

[7] ANONYMOUS, *Next slide please*, Nature, 272 (1978), p. 743.

[8] DON ASLETT, *Is There a Speech Inside You?*, Writer's Digest Books, Cincinnati, Ohio, 1989. ISBN 0-89879-361-0.

[9] DAVID H. BAILEY, *Twelve ways to fool the masses when giving performance results on parallel computers*, Supercomputer Rev., August 1991, pp. 54–55. Also in Supercomputer, Sept. 1991, pp. 4–7.

[10] SHERIDAN BAKER, *The Practical Stylist*, sixth ed., Harper and Row, New York, 1985. ISBN 0-06-040439-6.

[11] ROBERT BARRASS, *Scientists Must Write: A Guide to Better Writing for Scientists, Engineers and Students*, Chapman and Hall, London, 1978. ISBN 0-412-15430-7.

[12] ROBERT BARRASS, *Students Must Write: A Guide to Better Writing in Course Work and Examinations*, Methuen, London, 1982. ISBN 0-416-33620-5.

[13] FLOYD K. BASKETTE, JACK Z. SISSORS, AND BRIAN S. BROOKS, *The Art of Editing*, fifth ed., Macmillan, New York, 1992.

[14] DAVID F. BEER, ED., *Writing and Speaking in the Technology Professions: A Practical Guide*, IEEE Press, New York, 1992. ISBN 0-87942-284-X.

[15] ALBERT H. BEILER, *Recreations in the Theory of Numbers: The Queen of Mathematics Entertains*, Dover, New York, 1966. ISBN 0-486-21096-0.

[16] JON L. BENTLEY, *Programming Pearls*, Addison-Wesley, Reading, Massachusetts, 1986. ISBN 0-201-10331-1.

[17] ――――, *More Programming Pearls: Confessions of a Coder*, Addison-Wesley, Reading, Massachusetts, 1988. ISBN 0-201-11889-0.

[18] JON L. BENTLEY AND BRIAN W. KERNIGHAN, *Tools for printing indexes*, Electronic Publishing, 1 (1988), pp. 3–17. Also Computer Science Technical Report No. 128, AT&T Bell Laboratories, Murray Hill, New Jersey, October 1986.

[19] A. BERMAN AND R. J. PLEMMONS, *Nonnegative Matrices in the Mathematical Sciences*, Academic Press, New York, 1979.

[20] ANDRÉ BERNARD, ED., *Rotten Rejections: A Literary Companion*, Penguin, New York, 1990. ISBN 0-14-014786-1.

[21] THEODORE M. BERNSTEIN, *Miss Thistlebottom's Hobgoblins: The Careful Writer's Guide to the Taboos, Bugbears and Outmoded Rules of English Usage*, Farrar, Straus and Giroux, New York, 1971. ISBN 0-671-50404-5. Reprinted by Simon and Schuster, New York, 1984.

[22] CICELY BERRY, *Your Voice and How to Use It Successfully*, Harrap, London, 1975. ISBN 0-245-52886-5.

[23] AMBROSE BIERCE, *Write it Right: A Little Blacklist of Literary Faults*, The Union Library Association, New York, 1937. Reprinted in [21, pp. 209–253].

[24] AKE. BJÖRCK, *Least squares methods*, in Handbook of Numerical Analysis, P. G. Ciarlet and J. L. Lions, eds., vol. I: Finite Difference Methods— Solution of Equations in $\mathbb{R}^n$, North-Holland, Amsterdam, 1990.

[25] LENNART BJÖRCK, MICHAEL KNIGHT, AND ELEANOR WIKBORG, *The Writing Process: Composition Writing for University Students*, second ed., Studentlitteratur, Lund, Sweden and Chartwell Bratt, Bromley, Kent, England, 1990. ISBN 91 44 28222 2 and 0 86238 300 5.

[26] RALPH P. BOAS, *Can we make mathematics intelligible?*, Amer. Math. Monthly, 88 (1981), pp. 727–731.

[27] VERNON BOOTH, *Communicating in Science: Writing and Speaking*, Cambridge University Press, Cambridge, 1984. ISBN 0-521-27771-X.

[28] L. A. BRANKIN AND A. M. MUMFORD, *File formats for computer graphics: Unraveling the confusion*, Technical Report TR1/92, NAG Ltd., Oxford, June 1992.

[29] BILL BRYSON, *The Penguin Dictionary of Troublesome Words*, second ed., Penguin, London, 1987. ISBN 0-14-051200-4.

[30] ———, *Mother Tongue: English and How It Got That Way*, Avon Books, New York, 1990. ISBN 0-380-71543-0.

[31] ROBERT BURCHFIELD, *Unlocking the English Language*, Faber and Faber, London, 1989.

[32] DAVID M. BURTON, *Elementary Number Theory*, Allyn and Bacon, Boston, 1980. ISBN 0-205-06978-9.

[33] JUDITH BUTCHER, *Copy-Editing: The Cambridge Handbook for Editors, Authors and Publishers*, third ed., Cambridge University Press, Cambridge, 1992. ISBN 0 521 40074 0.

[34] BILL BUZBEE, *Poisson's equation revisited*, Current Contents, No. 36, Sept. 7, 1992, p. 8.

[35] GEORGE D. BYRNE, *How to improve technical presentations*, SIAM News, 20 (1987), pp. 10–11.

[36] FLORIAN CAJORI, *A History of Mathematical Notations. Volume II. Notations Mainly in Higher Mathematics*, The Open Court Publishing Company, Chicago, 1929.

[37] JAMES CALNAN AND ANDRAS BARABAS, *Speaking at Medical Meetings: A Practical Guide*, second ed., William Heinemann Medical, London, 1981. ISBN 0 433 05001 2.

[38] DEBRA CAMERON AND BILL ROSENBLATT, *Learning GNU Emacs*, O'Reilly & Associates, Sebastopol, CA, 1991. ISBN 0-937175-84-6.

[39] G. V. CAREY, *Mind the Stop: A Brief Guide to Punctuation with a Note on Proof-Correction*, second ed., Penguin, London, 1958. ISBN 0-14-051072-9. Reprinted 1976.

[40] CHAMBERS, *Chambers Concise Dictionary*, Edinburgh, 1991.

[41] T. W. CHAUNDY, P. R. BARRETT, AND C. BATEY, *The Printing of Mathematics*, Oxford University Press, London, 1954.

[42] PEHONG CHEN AND MICHAEL A. HARRISON, *Index preparation and processing*, Software—Practice and Experience, 18 (1988), pp. 897–915.

[43] LORINDA L. CHERRY, *Writing tools*, IEEE Trans. Comm., COM-30 (1982), pp. 100–105.

[44] COLLINS, *Collins Cobuild English Language Dictionary*, London, 1987. ISBN 0 00375021 3.

[45] ———, *Collins English Dictionary*, third ed., London, 1991.

[46] BRUCE M. COOPER, *Writing Technical Reports*, Penguin, London, 1964. ISBN 0-14-020676-0. Reprinted 1986.

[47] H. CROWDER, R. S. DEMBO, AND J. M. MULVEY, *On reporting computational experiments with mathematical software*, ACM Trans. Math. Software, 5 (1979), pp. 193–203.

[48] DAVID CRYSTAL, *The English Language*, Penguin, London, 1988. ISBN 0-14-022730-X.

[49] JOHN DAINTITH AND R. D. NELSON, EDS., *The Penguin Dictionary of Mathematics*, Penguin, London, 1989. ISBN 0-14-051119-9.

[50] P. J. DAVIS, *Fidelity in mathematical discourse: Is one and one really two?*, Amer. Math. Monthly, 79 (1972), pp. 252–263.

[51] ROBERT A. DAY, *How to Write and Publish a Scientific Paper*, third ed., Cambridge University Press, Cambridge, and Oryx Press, Phoenix, Arizona, 1988. ISBN 0 521 36760 3.

[52] ———, *Scientific English: A Guide for Scientists and Other Professionals*, Oryx Press, Phoenix, Arizona, 1992. ISBN 0-89774-722-4.

[53] BERNARD DIXON, *Sciwrite*, Chemistry in Britain, 9 (1973), pp. 70–72.

[54] JACK J. DONGARRA AND ERIC GROSSE, *Distribution of mathematical software via electronic mail*, Comm. ACM, 30 (1987), pp. 403–407.

[55] JACK J. DONGARRA AND BILL ROSENER, *NA-NET: Numerical analysis NET*, Technical Report CS-91-146, Department of Computer Science, University of Tennessee, Knoxville, Sept. 1991.

[56] SUSAN DRESSEL AND JOE CHEW, *Authenticity beats eloquence*, IEEE Trans. Prof. Comm., PC-30 (1987), pp. 82–83.

[57] HANS F. EBEL, CLAUS BLIEFERT, AND WILLIAM E. RUSSEY, *The Art of Scientific Writing: From Student Reports to Professional Publications in Chemistry and Related Fields*, VCH Publishers, New York, 1987. ISBN 0-89573-645-4.

[58] ANNE EISENBERG, *Guide to Technical Editing: Discussion, Dictionary, and Exercises*, Oxford University Press, New York, 1992. ISBN 0-19-506306-6.

[59] J. R. EWER AND G. LATORRE, *A Course in Basic Scientific English*, Longman, Harlow, Essex, 1969. ISBN 0 582 52009 6.

[60] HARLEY FLANDERS, *Manual for Monthly authors*, Amer. Math. Monthly, 78 (1971), pp. 1–10.

[61] RUDOLF FLESCH, *The Art of Plain Talk*, Harper and Brothers, New York, 1946.

[62] BERNARD K. FORSCHER, *The role of the referee*, Scholarly Publishing, 11 (1980), pp. 165–169.

[63] H. W. FOWLER, *A Dictionary of Modern English Usage*, second ed., Oxford University Press, Oxford, 1968. Revised by Sir Ernest Gowers.

[64] H. W. FOWLER AND F. G. FOWLER, *The King's English*, third ed., Oxford University Press, Oxford, 1931. ISBN 0-19-881330-9. Reprinted 1990.

[65] DANIEL H. FREEMAN, JR., MARÍA ELENA GONZÁLEZ, DAVID C. HOAGLIN, AND BETH A. KILSS, *Presenting statistical papers*, Amer. Statist., 37 (1983), pp. 106–110.

[66] MATTHEW P. GAFFNEY AND LYNN ARTHUR STEEN, *Annotated Bibliography of Expository Writing in the Mathematical Sciences*, Mathematical Association of America, Washington, D.C., 1976. ISBN 0-88385-422-8.

[67] EUGENE GARFIELD, *"Science Citation Index"—A new dimension in indexing*, Science, 144 (1964), pp. 649–654.

[68] ——, *Citation analysis as a tool in journal evaluation*, Science, 178 (1972), pp. 471–479.

[69] ——, *Significant journals of science*, Nature, 264 (1976), pp. 609–615.

[70] ——, *Citation Indexing: Its Theory and Application in Science, Technology, and Humanities*, John Wiley, New York, 1979.

[71] ——, *Journal citation studies. 36. Pure and applied mathematics journals: What they cite and vice versa*, Current Contents, No. 15, March 8, 1982, pp. 5–10.

[72] ——, *The 100 most-cited papers ever and how we select citation classics*, Current Contents, No. 23, June 4, 1984, pp. 5–11.

[73] ——, *The Awards of Science and Other Essays*, Essays of an Information Scientist: 1984, ISI Press, Philadelphia, 1985.

[74] EUGENE GARFIELD, *Journal editors awaken to the impact of citation errors. How we control them at ISI*, Current Contents, No. 41, October 8, 1990, pp. 5–13.

[75] ———, *The most-cited papers of all time, SCI 1945–1988. Part 1A. The SCI top 100—will the Lowry method ever be obliterated?*, Current Contents, No. 7, February 12, 1990, pp. 3–14.

[76] ———, *The most-cited papers of all time, SCI 1945–1988. Part 1B. Superstars new to the SCI top 100*, Current Contents, No. 8, February 19, 1990, pp. 3–13.

[77] ———, *The most-cited papers of all time, SCI 1945–1988. Part 3. Another 100 from the Citation Classics Hall of Fame*, Current Contents, No. 34, August 20, 1990, pp. 3–13.

[78] ———, *How to Use the Science Citation Index (SCI)*, in SCI Science Citation Index 1991. Guide and List of Source Publications, ISI Press, Philadelphia, 1991, pp. 26–33. Reprinted from Current Contents, No. 9, February 28, 1983, pp. 5–14.

[79] ROBERT V. GARVER, *Presenting the peer paper*, IEEE Trans. Prof. Comm., PC-23 (1980), pp. 18–22.

[80] C. WILLIAM GEAR, *Inside SIAM's mysterious journal publication process*, SIAM News, 24 (2) (1991), p. 6.

[81] LEONARD GILLMAN, *Writing Mathematics Well: A Manual for Authors*, The Mathematical Association of America, Washington, D.C., 1987. ISBN 0-88385-443-0.

[82] ———, *Paul Halmos's expository writing*, in Paul Halmos: Celebrating 50 Years of Mathematics, John H. Ewing and F. W. Gehring, eds., Springer-Verlag, Berlin, 1991, pp. 33–48.

[83] LEON J. GLESER, *Some notes on refereeing*, Amer. Statist., 40 (1986), pp. 310–312.

[84] GENE H. GOLUB AND CHARLES F. VAN LOAN, *Matrix Computations*, second ed., Johns Hopkins University Press, Baltimore, Maryland, 1989. ISBN 0-8018-3739-1.

[85] CALVIN R. GOULD, *The overhead projector*, IEEE Trans. Prof. Comm., PC-15 (1972), pp. 2–6.

[86] ———, *Visual aids—how to make them positively legible*, IEEE Trans. Prof. Comm., PC-16 (1973), pp. 35–38.

[87] SIR ERNEST GOWERS, *The Complete Plain Words*, third ed., Penguin, London, 1986. ISBN 0-14-051199-7. Revised by Sidney Greenbaum and Janet Whitcut.

[88] RONALD L. GRAHAM, DONALD E. KNUTH, AND OREN PATASHNIK, *Concrete Mathematics: A Foundation for Computer Science*, Addison-Wesley, Reading, Massachusetts, 1989. ISBN 0-201-14236-8.

[89] MARTIN W. GREGORY, *The infectiousness of pompous prose*, Nature, 360 (1992), pp. 11–12.

[90] R. GRONE, C. R. JOHNSON, E. M. SÁ, AND H. WOLKOWICZ, *Normal matrices*, Linear Algebra Appl., 87 (1987), pp. 213–225.

[91] PAUL R. HALMOS, *How to write mathematics*, Enseign. Math., 16 (1970), pp. 123–152. Reprinted in [192] and [182].

[92] ——, *Finite-Dimensional Vector Spaces*, Springer-Verlag, New York, 1974.

[93] ——, *How to talk mathematics*, Notices Amer. Math. Soc., 21 (1974), pp. 155–158. Reprinted in [182].

[94] ——, *What to publish*, Amer. Math. Monthly, 82 (1975), pp. 14–17. Reprinted in [182].

[95] ——, *A Hilbert Space Problem Book*, second ed., Springer-Verlag, Berlin, 1982.

[96] ——, *Think it gooder*, Math. Intelligencer, 4 (1982), pp. 20–21.

[97] ——, *I Want to Be a Mathematician: An Automathography in Three Parts*, Springer-Verlag, New York, 1985. ISBN 0-88385-445-7.

[98] ——, *Some books of auld lang syne*, in A Century of Mathematics in America, Part I, Peter Duren, Richard A. Askey, and Uta C. Merzbach, eds., American Mathematical Society, Providence, Rhode Island, 1988, pp. 131–174. ISBN 0-8218-0124-4.

[99] HORACE HART, *Hart's Rules for Compositors and Readers at the University Press Oxford*, thirty-ninth ed., Oxford University Press, Oxford, 1983. ISBN 0-19-212983-X.

[100] JAMES HARTLEY, *Eighty ways of improving instructional text*, IEEE Trans. Prof. Comm., PC-24 (1981), pp. 17–27.

[101] ——, *The role of colleagues and text-editing programs in improving text*, IEEE Trans. Prof. Comm., PC-27 (1984), pp. 42–44.

[102] ——, *Designing Instructional Text*, second ed., Kogan Page, London, 1985. ISBN 0 85038 943 7.

[103] EDWARD F. HARTREE, *Ethics for authors: A case history of acrosin*, Perspect. in Biol. and Med., 20 (1976), pp. 82–91.

[104] J. B. HEATON AND N. D. TURTON, *Longman Dictionary of Common Errors*, Longman, Harlow, Essex, 1987. ISBN 0 582 96410 5.

[105] A. J. HERBERT, *The Structure of Technical English*, Longman, Harlow, Essex, 1965. ISBN 0 582 52523 3.

[106] DESMOND J. HIGHAM, *More commandments of good writing*, Manuscript, Department of Mathematics and Computer Science, University of Dundee, Nov. 1992.

[107] NICHOLAS J. HIGHAM, *Algorithm 694: A collection of test matrices in MATLAB*, ACM Trans. Math. Software, 17 (1991), pp. 289–305.

[108] PHILIP J. HILLS, ED., *Publish or Perish*, Peter Francis Publishers, Berrycroft, Cambs., 1987. ISBN 1-8701-6700-7.

[109] HOUGHTON MIFFLIN, *The American Heritage Dictionary of the English Language*, 2nd college ed., Boston, 1982. 1985 printing.

[110] ALSTON S. HOUSEHOLDER, *The Theory of Matrices in Numerical Analysis*, Blaisdell, New York, 1964. Reprinted by Dover, New York, 1975. ISBN 0-486-61781-5.

[111] KENNETH E. IVERSON, *A Programming Language*, Wiley, New York, 1962.

[112] R. H. F. JACKSON, P. T. BOGGS, S. G. NASH, AND S. POWELL, *Guidelines for reporting results of computational experiments. Report of the ad hoc committee*, Math. Programming, 49 (1991), pp. 413–425.

[113] J. L. KELLEY, *Writing mathematics*, in Paul Halmos: Celebrating 50 Years of Mathematics, John H. Ewing and F. W. Gehring, eds., Springer-Verlag, Berlin, 1991, pp. 91–96.

[114] KEVIN KELLY, ED., *SIGNAL: Communication Tools for the Information Age*, Harmony Books, Crown Publishers, 225 Park Ave. South, New York, NY 10003, 1988. ISBN 0-517-57084-X.

[115] PETER KENNY, *A Handbook of Public Speaking for Scientists and Engineers*, Adam Hilger, Bristol, 1982. ISBN 0-85274-553-2.

[116] G. A. KERKUT, *Choosing a title for a paper*, Comp. Biochem. Physiol., 47A (1983), p. 1.

[117] BRIAN W. KERNIGHAN AND LORINDA L. CHERRY, *A system for typesetting mathematics*, Comm. ACM, 18 (1975), pp. 151–157.

[118] LESTER S. KING, *Medical writing number 7: The opening sentence*, J. Amer. Medical Assoc., 202 (1967), pp. 535–536.

[119] JOHN KIRKMAN, *Good Style: Writing for Science and Technology*, E & FN Spon (Chapman and Hall), London, 1992. ISBN 0 419 17190 8.

[120] GEORGE R. KLARE, *The Measurement of Readability*, Iowa State University Press, Ames, Iowa, 1963.

[121] DONALD E. KNUTH, *The Art of Computer Programming*, Addison-Wesley, Reading, Massachusetts, 1973–1981. Three volumes.

[122] ———, *Mathematical typography*, Bull. Amer. Math. Soc. (New Series), 1 (1979), pp. 337–372.

[123] ———, *The Art of Computer Programming, Volume 2, Seminumerical Algorithms*, second ed., Addison-Wesley, Reading, Massachusetts, 1981. ISBN 0-201-03822-6.

[124] ———, *The TEXbook*, Addison-Wesley, Reading, Massachusetts, 1986. ISBN 0-201-13448-9.

[125] ———, *The METAFONT Book*, Addison-Wesley, Reading, Massachusetts, 1986. ISBN 0-201-13444-6.

[126] ———, *3:16 Bible Texts Illuminated*, A-R Editions, 801 Deming Way, Madison, Wisconsin, 1991. ISBN 0 89579 252 4.

[127] ———, *Two notes on notation*, Amer. Math. Monthly, 99 (1992), pp. 403–422.

[128] DONALD E. KNUTH, TRACY LARRABEE, AND PAUL M. ROBERTS, *Mathematical Writing*, MAA Notes Number 14, The Mathematical Association of America, Washington, D.C., 1989. ISBN 0-88385-063-X. Also Report STAN-CS-88-1193, Department of Computer Science, Stanford University, January 1988.

[129] KODAK LIMITED, *Let's stamp out awful lecture slides*, Kodak publication S-22(H), Apr. 1979.

[130] ED KROL, *The Whole Internet User's Guide & Catalog*, O'Reilly & Associates, Sebastopol, CA, 1992. ISBN 1-56592-025-2.

[131] LESLIE LAMPORT, *LATEX: A Document Preparation System*, Addison-Wesley, Reading, Massachusetts, 1986. ISBN 0-201-15790-X.

[132] ———, *Document production: Visual or logical?*, Notices Amer. Math. Soc., 34 (1987), pp. 621–624.

[133] ———, *Makeindex: An index processor for LATEX*. Documentation distributed with MakeIndex program, 17 Feb. 1987.

[134] TRACY LAQUEY AND JEANNE C. RYER, *The Internet Companion: A Beginner's Guide to Global Networking*, Addison-Wesley, Reading, Massachusetts, 1993. ISBN 0-201-62224-6.

[135] STEPHEN LOCK, ED., *How to Do It*, second ed., British Medical Association, London, 1985. ISBN 0-7279-0186-9.

[136] LONGMAN, *Longman Dictionary of Contemporary English*, second ed., Harlow, Essex, 1987. ISBN 0 582 84222 0.

[137] ——, *Longman Dictionary of the English Language*, new ed., Harlow, Essex, 1991. ISBN 0 582 07038 4.

[138] BETH LUEY, *Handbook for Academic Authors*, Cambridge University Press, Cambridge, 1987. ISBN 0-521-33682-1.

[139] NINA H. MACDONALD, LAWRENCE T. FRASE, PATRICIA S. GINGRICH, AND STACEY A. KEENAN, *The writer's workbench: Computer aids for text analysis*, IEEE Trans. Comm., COM-30 (1982), pp. 105–110.

[140] A. J. MACGREGOR, *Graphics simplified: Charts and graphs*, Scholarly Publishing, 8 (1977), pp. 151–164.

[141] ——, *Graphics simplified: Preparing charts and graphs*, Scholarly Publishing, 8 (1977), pp. 257–274.

[142] DONALD S. MACQUEEN, *Using Numbers in English: A Reference Guide for Swedish Speakers Including Basic Terminology for Describing Graphs*, Studentlitteratur, Lund, Sweden and Chartwell Bratt, Bromley, Kent, England, 1990. ISBN 91 4431921 5 and 0 862382645.

[143] JOHN MADDOX, *Must science be impenetrable?*, Nature, 305 (1983), pp. 477–478.

[144] FRANK T. MANHEIM, *The scientific referee*, IEEE Trans. Prof. Comm., PC-18 (1975), pp. 190–195.

[145] GLENDA M. MCCLURE, *Readability formulas: Useful or useless?*, IEEE Trans. Prof. Comm., PC-30 (1987), pp. 12–15.

[146] M. DOUGLAS MCILROY, *Development of a spelling list*, IEEE Trans. Comm., COM-30 (1982), pp. 91–99.

[147] MERRIAM-WEBSTER, *Webster's Ninth New Collegiate Dictionary*, Springfield, Massachusetts, 1991. ISBN 0-87779-509-6.

[148] JAMES A. MICHENER, *James A. Michener's Writer's Handbook: Explorations in Writing and Publishing*, Random House, New York, 1992. ISBN 0-679-74126-7.

[149] JOAN P. MITCHELL, *The New Writer: Techniques for Writing Well with a Computer*, Microsoft Press, Redmond, Washington, 1987. ISBN 1-55615-029-6.

[150] MAEVE O'CONNOR, *Editing Scientific Books and Journals*, Pitman Medical, Tunbridge Wells, Kent, 1978. ISBN 0-27279517-8.

[151] ———, *How to Copyedit Scientific Books and Journals*, ISI Press, Philadelphia, 1986. ISBN 0-89495-064-9.

[152] ———, *Writing Successfully in Science*, Chapman and Hall, London, 1991. ISBN 0 412 446308.

[153] MAEVE O'CONNOR AND F. PETER WOODFORD, *Writing Scientific Papers in English*, Pitman Medical, Tunbridge Wells, Kent, 1977. ISBN 0-272-79515-1.

[154] D. P. O'LEARY, G. W. STEWART, AND J. S. VANDERGRAFT, *Estimating the largest eigenvalue of a positive definite matrix*, Math. Comp., 33 (1979), pp. 1289–1292.

[155] J. M. ORTEGA, R. G. VOIGT, AND C. H. ROMINE, *A bibliography on parallel and vector numerical algorithms*, in Parallel Algorithms for Matrix Computations, Society for Industrial and Applied Mathematics, Philadelphia, 1990, pp. 125–197.

[156] OXFORD UNIVERSITY PRESS, *Oxford Advanced Learner's Dictionary of Current English*, fourth ed., 1989. ISBN 0 19431141 4.

[157] ———, *The Oxford English Dictionary*, second ed., 1989.

[158] ———, *The Concise Oxford Dictionary of Current English*, eighth ed., 1990. ISBN 0-19-861200-1.

[159] RICHARD S. PALAIS, *An introduction to TEX and AMS-TEX*, Notices Amer. Math. Soc., 33 (1986), pp. 303–308.

[160] BERESFORD N. PARLETT, *The Symmetric Eigenvalue Problem*, Prentice Hall, Englewood Cliffs, New Jersey, 1980. ISBN 0-13-880047-2.

[161] ERIC PARTRIDGE, *Usage and Abusage: A Guide to Good English*, Penguin, London, 1973. ISBN 0-14-051024-9.

[162] OREN PATASHNIK, *BIBTEXing*. Documentation for general BIBTEX users. File `btxdoc.tex` in the standard distribution, 8 Feb. 1988.

[163] JAN A. PECHENIK, *A Short Guide to Writing about Biology*, HarperCollins, New York, 1987. ISBN 0-673-39232-5.

[164] JOHN E. PEMBERTON, *How to Find Out in Mathematics*, second ed., Pergamon, London, 1969.

[165] CAROL ROSENBLUM PERRY, *The Fine Art of Technical Writing*, Blue Heron Publishing, Hillsboro, Oregon, 1991. ISBN 0-936085-24-X.

[166] IVARS PETERSON, *Searching for new mathematics*, SIAM Review, 33 (1991), pp. 37–42.

[167] JAMES L. PETERSON, *Computer Programs for Spelling Correction: An Experiment in Program Design*, no. 96 in Lecture Notes in Computer Science, Springer-Verlag, Berlin, 1980.

[168] ———, *A note on undetected typing errors*, Comm. ACM, 29 (1986), pp. 633–637.

[169] GEORGE PIRANIAN, *Say it better*, Math. Intelligencer, 4 (1982), pp. 17–19.

[170] GEORGE POLYA, *How to Solve It: A New Aspect of Mathematical Method*, second ed., Doubleday, New York, 1957.

[171] SIMEON POTTER, *Our Language*, Penguin, London, 1976.

[172] PRENTICE HALL, *Webster's New World Dictionary of American English*, third college ed., Englewood Cliffs, New Jersey, 1988.

[173] W. M. PRIESTLEY, *Paul Halmos: Words more than numbers*, in Paul Halmos: Celebrating 50 Years of Mathematics, John H. Ewing and F. W. Gehring, eds., Springer-Verlag, Berlin, 1991, pp. 49–69.

[174] D. A. PYKE, *Referee a paper*, in How to Do It, second ed., Stephen Lock, ed., British Medical Association, London, 1985, pp. 215–219.

[175] RANDOLPH QUIRK AND GABRIELE STEIN, *English in Use*, Longman, Harlow, Essex, 1990. ISBN 0 582 06613 1.

[176] RANDOM HOUSE, *The Random House College Dictionary*, revised ed., New York, 1988.

[177] ERVIN Y. RODIN, *Speed of publication—an editorial*, Comput. Math. Appl., 24 (1992), pp. 1–2.

[178] CHARLES G. ROLAND, *Thoughts about medical writing XXXVII. Verify your references*, Anesthesia and Analgesia ... Current Researches, 55 (1976), pp. 717–718.

[179] RICHARD RUBINSTEIN, *Digital Typography: An Introduction to Type and Composition for Computer System Design*, Addison-Wesley, Reading, Massachusetts, 1988. ISBN 0-201-17633-5.

[180] STEPHAN M. RUDOLFER AND PETER C. WATSON, *Table errata*, Math. Comp., 59 (1992), p. 727.

[181] WILLIAM SAFIRE, *Fumblerules: A Lighthearted Guide to Grammar and Good Usage*, Dell Publishing, New York, 1990. ISBN 0-440-21010-0.

[182] DONALD E. SARASON AND LEONARD GILLMAN, EDS., *P. R. Halmos. Selecta: Expository Writing*, Springer-Verlag, New York, 1983. ISBN 0-387-90756-4.

[183] DAVID LOUIS SCHWARTZ, *How to be a published mathematician without trying harder than necessary*, in The Journal of Irreproducible Results: Selected Papers, third ed., George H. Scherr, ed., 1986, p. 205.

[184] MARJORIE E. SKILLIN, ROBERT M. GAY, AND OTHER AUTHORITIES, *Words into Type*, third ed., Prentice-Hall, Englewood Cliffs, New Jersey, 1974.

[185] MICHAEL D. SPIVAK, *The Joy of T_EX: A Gourmet Guide to Typesetting with the A_MS-T_EX Macro Package*, second ed., American Mathematical Society, Providence, Rhode Island, 1990.

[186] ——, *LA_MS-T_EX*, The T_EXplorators Corporation, Houston, Texas, 1991.

[187] ELSIE MYERS STAINTON, *A bag for editors*, Scholarly Publishing, 8 (1977), pp. 111–119.

[188] ——, *The uses of dictionaries*, Scholarly Publishing, 11 (1980), pp. 229–241.

[189] ——, *The Fine Art of Copyediting*, Columbia University Press, New York, 1991. ISBN 0-231-06961-8.

[190] RICHARD STALLMAN, *GNU Emacs Manual, Emacs Version 18*, fifth ed., Free Software Foundation, 675 Mass Ave, Cambridge, Massachusetts 02139, USA, Oct. 1986.

[191] NORMAN E. STEENROD, *How to write mathematics*, in [192], 1973, pp. 1–17.

[192] NORMAN E. STEENROD, PAUL R. HALMOS, MENAHEM M. SCHIFFER, AND JEAN A. DIEUDONNÉ, *How to Write Mathematics*, American Mathematical Society, Providence, Rhode Island, 1973.

[193] ANDREW STERRETT, ED., *Using Writing to Teach Mathematics*, MAA Notes Number 16, The Mathematical Association of America, Washington, D.C., 1990. ISBN 0-88385-066-4.

[194] HANS J. STETTER, *Analysis of Discretization Methods for Ordinary Differential Equations*, Springer-Verlag, Berlin, 1973.

[195] G. W. STEWART, *Introduction to Matrix Computations*, Academic Press, New York, 1973. ISBN 0-12-670350-7.

[196] G. W. STEWART, *FTPing technical reports*, NA Digest, 91 (39), 1991. Electronic mail magazine: na.help@na-net.ornl.gov.

[197] GILBERT STRANG, *Introduction to Applied Mathematics*, Wellesley-Cambridge Press, Wellesley, Massachusetts, 1986. ISBN 0-9614088-0-4.

[198] WILLIAM STRUNK, JR. AND E. B. WHITE, *The Elements of Style*, third ed., Macmillan, New York, 1979. ISBN 0-02-418200-1.

[199] JOHN SWALES, *Writing Scientific English*, Thomas Nelson, London, 1971.

[200] ———, *Episodes in ESP: A Source and Reference Book on the Development of English for Science and Technology*, Prentice Hall International, Hemel Hempstead, Hampshire, 1988. ISBN 0-13-283383-2.

[201] MICHAEL SWAN, *Practical English Usage*, Oxford University Press, Oxford, 1980. ISBN 0-19-431185-6.

[202] ELLEN SWANSON, *Mathematics into Type: Copy Editing and Proofreading of Mathematics for Editorial Assistants and Authors*, revised ed., American Mathematical Society, Providence, Rhode Island, 1979. ISBN 0-8218-0053-1.

[203] JUDITH A. TARUTZ, *Technical Editing: The Practical Guide for Editors and Writers*, Addison-Wesley, Reading, Massachusetts, 1992. ISBN 0-201-56356-8.

[204] ROBERT C. THOMPSON, *Author vs. referee: A case history for middle level mathematicians*, Amer. Math. Monthly, 90 (1983), pp. 661–668.

[205] MARTIN TOMPA, *Figures of merit*, Research Report RC 14211 (#63576), IBM Thomas J. Watson Research Center, Yorktown Heights, New York, Nov. 1988.

[206] EDWARD R. TUFTE, *The Visual Display of Quantitative Information*, Graphics Press, Cheshire, Connecticut, 1983.

[207] KATE L. TURABIAN, *A Manual for Writers of Term Papers, Theses, and Dissertations*, fifth ed., The University of Chicago Press, Chicago and London, 1987. ISBN 0 226 81625 7.

[208] CHRISTOPHER TURK, *Effective Speaking: Communicating in Speech*, E & FN Spon (Chapman and Hall), London, 1985. ISBN 0 419 13030 6.

[209] CHRISTOPHER TURK AND JOHN KIRKMAN, *Effective Writing: Improving Scientific, Technical and Business Communication*, second ed., E & FN Spon (Chapman and Hall), London, 1989. ISBN 0 419 14660 1.

[210] BARRY T. TURNER, *Effective Technical Writing and Speaking*, second ed., Business Books, London, 1974. ISBN 0 220 66227 4.

[211] ADRIAN UNDERHILL, *Use Your Dictionary: A Practice Book for Users of Oxford Advanced Learner's Dictionary of Current English and Oxford Student's Dictionary of Current English*, Oxford University Press, Oxford, 1980. ISBN 0 19 431104 X.

[212] UNIVERSITY OF CHICAGO PRESS, *The Chicago Manual of Style*, thirteenth ed., Chicago and London, 1982. ISBN 0-226-10390-0.

[213] MARY-CLAIRE VAN LEUNEN, *A Handbook for Scholars*, revised ed., Oxford University Press, New York, 1992. ISBN 0-19-506954-4. First edition Knopf, 1978.

[214] CHARLES F. VAN LOAN, *Computational Frameworks for the Fast Fourier Transform*, Society for Industrial and Applied Mathematics, Philadelphia, 1992. ISBN 0-89871-285-8.

[215] KEITH WATERHOUSE, *On Newspaper Style*, Viking, London, 1989. ISBN 0-670-82626-X.

[216] ——, *English our English (and How to Sing It)*, Viking, London, 1991. ISBN 0-670-83269-3.

[217] DAVID S. WATKINS, *Fundamentals of Matrix Computations*, John Wiley, New York, 1991. ISBN 0-471-54601-1.

[218] J. D. WATSON AND F. H. C. CRICK, *Molecular structure of nucleic acids: A structure for deoxyribose nucleic acid*, Nature, 171 (1953), pp. 737–738. April 25.

[219] HERBERT S. WILF, *TEX: A non-review*, Amer. Math. Monthly, 93 (1986), pp. 309–315.

[220] J. H. WILKINSON, *The Algebraic Eigenvalue Problem*, Oxford University Press, Oxford, 1965. ISBN 0-19-853403-5.

[221] FREDERICK T. WOOD, ROGER H. FLAVELL, AND LINDA M. FLAVELL, *Current English Usage*, Macmillan, London, 1989. ISBN 0-333-27840-2.

[222] F. PETER WOODFORD, *Sounder thinking through clearer writing*, Science, 156 (1967), pp. 743–745.

[223] WILLIAM ZINSSER, *Writing with a Word Processor*, Harper and Row, New York, 1983. ISBN 0-06-091060-7.

[224] ——, *Writing to Learn*, Harper and Row, New York, 1988. ISBN 0-06-091576-5.

[225] ——, *On Writing Well: An Informal Guide to Writing Nonfiction*, fourth ed., Harper-Collins, New York, 1990. ISBN 0-06-272027-9.

Index